Encyclopedia of Genetic Diversity in Plants: Advanced Concepts

Volume II

Encyclopedia of Genetic Diversity in Plants: Advanced Concepts Volume II

Edited by **Harvey Parker**

New York

Published by Callisto Reference,
106 Park Avenue, Suite 200,
New York, NY 10016, USA
www.callistoreference.com

Encyclopedia of Genetic Diversity in Plants: Advanced Concepts
Volume II
Edited by Harvey Parker

International Standard Book Number: 978-1-63239-253-4 (Hardback)

Printed in the United States of America.

Contents

Permissions

List of Contributors

Preface

This book was inspired by the evolution of our times; to answer the curiosity of inquisitive minds. Many developments have occurred across the globe in the recent past which has transformed the progress in the field.

An examination of genetic diversity in plants on the basis of research-focused information has been described in this book. Genetic diversity is of central importance to the continuity of a species as it presents with the required adaptation to the current biotic and abiotic ecological circumstances, and enables modifications in the genetic composition to cope with changes in the environment. This book discusses the extent of genetic variation present in plant populations and conservation of germplasm. The motive of the book is to present the views of experts who are involved in the creation of fresh ideas and methods employed for the assessment of genetic diversity, often from various viewpoints. The book will be helpful to students, researchers and experts in the field.

This book was developed from a mere concept to drafts to chapters and finally compiled together as a complete text to benefit the readers across all nations. To ensure the quality of the content we instilled two significant steps in our procedure. The first was to appoint an editorial team that would verify the data and statistics provided in the book and also select the most appropriate and valuable contributions from the plentiful contributions we received from authors worldwide. The next step was to appoint an expert of the topic as the Editor-in-Chief, who would head the project and finally make the necessary amendments and modifications to make the text reader-friendly. I was then commissioned to examine all the material to present the topics in the most comprehensible and productive format.

I would like to take this opportunity to thank all the contributing authors who were supportive enough to contribute their time and knowledge to this project. I also wish to convey my regards to my family who have been extremely supportive during the entire project.

Editor

Conservation of Germplasms

A Brief Review of a Nearly Half a Century Wheat Quality Breeding in Bulgaria

Dobrinka Atanasova, Nikolay Tsenov and Ivan Todorov
Dobrudzha Agricultural Institute
General Toshevo
Bulgaria

1. Introduction

Wheat quality is a complex feature determined by the level of various indices and dependent on a number of factors. It is defined by different interactions involving glutenin and gliadin composition and abiotic stresses. The efforts of the wheat breeders are directed towards combining a high level of the indices in the new varieties with a stable expression of these indices in different years and environments.

The aim of this chapter is 1) to describe the glutenin and gliadin composition of Bulgarian winter bread wheat varieties, bred since the middle of 20th century; 2) to asses the quality indices of wheat varieties; to compare the differences between wheat quality groups as well as the different period of creating the varieties; 3) to determine the level of influence of various environmental factors on the quality indices of wheat; to follow the response of common winter wheat varieties to various combinations of growing conditions; to study the ability of the varieties to realize their genetic potential for quality under certain environments.

2. Genetic diversity of Bulgarian winter wheat varieties in relation to glutenin and gliadin compositions

Wheat storage proteins (gliadins and glutenins) are the main components of gluten which is the major determinant of end use quality. Glutenin proteins are divided into two groups: High and Low Molecular Weight Glutenin Subunits (HMW-GS and LMW-GS, respectively). HMW-GS are encoded by two type of genes (x and y) that are located on three loci (Glu-A1, Glu-B1 and Glu-D1) placed on the long arm of the group 1 chromosomes (Lawrence & Shepherd, 1981). LMW-GS are classically divided into B, C and D groups on the basis of molecular weight and isoelectric point (Jackson et al., 1983). They are coded by gene families located on the short arm of the group 1 chromosomes (Glu-A3, Glu-B3 and Glu-D3 loci). Gliadins are monomeric proteins and they are classified as α-, β-, γ- and ω-gliadins. Most of the α/ β-gliadins are encoded at the Gli-2 loci, located on the short arms of chromosomes 6A, 6B and 6D. Most of γ- and ω-gliadins are encoded at the Gli-1 loci, located on the short arms of chromosomes 1A, 1B and 1D (Masci et al., 2002). It has been reported a close linkage between the Glu-3 loci encoding LMW-GS and the Gli-1 loci for gliadins (Sing and Shepherd, 1988).

The bread making characteristics of wheat flour are closely related to the elasticity and extensibility of the gluten proteins. The seed storage proteins are associated with agriculturally significant traits and they are used in a legal protection of cultivars (Knoblochova & Galova, 2000). Many studies have been made in order to investigate the genetic diversity of the initial material (Atanasova et al., 2009b; Bradova, 2008; Bushehri et al., 2006; Li et al., 2005; Tabasum et al., 2011; Tohver, 2007; Tsenov et al., 2009).

Bulgarian wheat varieties, especially those developed at Dobrudzha Agricultural Institute - General Toshevo (DAI), are known to have good and very good end-use quality (Atanasova et al., 2008; Panayotov et al., 2004, Todorov, 2006; Tsenov et al., 2010a). Their glutenin composition has been investigated to various degrees (Atanasova et al., 2009b; Todorov et al., 2006; Tsenov et al., 2009).

In this review eighty-nine cultivars developed at Dobrudzha Agricultural Institute (DAI) – General Toshevo, Bulgaria, during 1962–2010, and nine cultivars of the Institute of Plant and Genetic Resources (IPGR) – Sadovo, Bulgaria were investigated. Some of these cultivars were heterogeneous and their biotypes were considered as separate varieties when determining the ratio of high- and low- molecular weight glutenin and gliadin subunits and spectra. The allelic frequency in loci Glu-1, Glu-3, and Gli-1 was studied and the genetic variability in each allele was calculated. The percent of observed spectra was determined by decades.

In high-molecular weight glutenins, highest variability was registered in locus Glu-B1 (fig. 1). Five alleles were identified – a, b, c, d, f, which expressed subunits 7, 7+8, 7+9, 6+8 and 13+16, as well as one untypical fraction pair 6+9 which was observed in a biotype of cultivar Skitiya. Allele c had highest frequency (65.4%), followed by allele b (17.7%) and allele d (10.8%). Three alleles were identified in locus Glu-A1: a, b, c, expressing the respective subunits 1 (24.6%), 2* (41.5) and N (33.8%). In locus Glu-D1 allele a (subunit 2+12) had frequency 27.7%, and allele d (subunit 5+10) - 69.2%. Biotypes with the untypical fraction pair 5+12 were observed in cultivar Levent.

In the low-molecular weight glutenins, locus Glu-B3 had highest variability, where 7 alleles were identified – b, d, f, g, h, i, j, followed by locus Glu-A3 with 5 alleles – b, c, d, e, f; in locus Glu-D3 two alleles were identified – a and c. Allele Glu-A3c (69.2%) had highest frequency, followed by allele Glu-A3e (18.5%). The other alleles had frequency less than 10 %. Allele Glu-B3b had frequency 47.7%, followed by Glu-B3j (13.8%), Glu-B3f – 13.1%, Glu-B3h – 10.0%.

The gliadin fraction composition of the investigated wheat accessions was more variable than the fraction composition of the high- and low-molecular weight glutenins. In each of all three gliadins loci 9 alleles were observed. In locus Gli-A1 the frequency of alleles Gli-A1b (40.8%), Gli-A1m (18.5%) and Gli-A1a (17.7%) was highest. In locus Gli-B1 the alleles with highest frequency were: Gli-B1b (47.7%), Gli-B1l (13.8%), Gli-B1g (13.1%) and Gli-B1d (10.0%). In locus Gli-D1 the alleles with highest frequency were Gli-D1b (56.2%), Gli-D1j (13.1%) and Gli-D1a (11.5%).

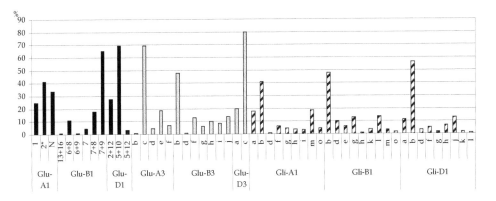

Fig. 1. The allelic frequency in loci *Glu-1*, *Glu-3*, and *Gli-1*

Twenty-four high-molecular weight, twenty-six low-molecular weight glutenin and fifty-seven gliadins spectra were identified in the investigated wheat accessions. The most frequent high-molecular weight spectra were 2* 7+9 5+10 (22.14%), N 7+9 5+10 (14.5%), 2* 7+9 2+12 (8.4%), 1 7+9 5+10 (8.4%) (table 1). Among the low-molecular weight glutenin spectra, the following had highest frequency: *c b c* (22.9%), *e b c* (12.2%), *c f c* (8.4%), *c h c* (6.9%), *c j c* (6.9%). In gliadins, most frequent were spectra *b b b* (12.2%), *m b b* (6.9%), *m b j* (6.1%).

Loci	Spectrum	%
Glu-1	2* 7+9 5+10	22.14
	N 7+9 5+10	14.5
	2* 7+9 2+12	8.4
	1 7+9 5+10	8.4
Glu-3	c b c	22.90
	e b c	12.21
	c f c	8.40
	c h c	6.9
	c j c	6.9
Gli-1	b b b	12.2
	m b b	6.9
	m b j	6.1

Table 1. The most frequent spectra in the *Glu-1*, *Glu-3* and *Gli-1* loci

The highest frequency of *Glu-1* spectra in the 70's was for N 7+9 5+10 (19.4%), 2* 7+9 2+12 (13.9%), 2* 7+9 5+10 (13.9%), N 7+9 2+12 (11.1%) (table 2). In the 80's these were 2* 7+9 5+10 (20.7%), 1 7+9 2+12 (10.3%), N 6+8 5+10 (10.3%), N 7+9 5+10 (10.3%). The most frequent in the 90's were the spectra 2* 7+9 5+10 (30.4%), N 7+9 5+10 (21.7%), 1 7+9 5+10 (17.4%), N 7+8 5+10 (13.0%). In the new century widespread were: 2* 7+9 5+10 (25.6%), 2* 7+8 5+10 (14.0%), 2* 7+9 2+12 (11.6%), 1 7+8 5+10 (9.3%), 1 7+9 5+10 (9.3%), N 7+9 5+10 (9.3%). The frequency of *Glu-3* and *Gli-1* spectra in the different decades is shown in the table.

Decade	Glu-1	Glu-3	Gli-1
	N 7+9 5+10 (19.4)*	c b c (33.3)	b b b (14.3)
	2* 7+9 2+12 (13.9)	c i c (16.7)	a b b (5.7)
70	2* 7+9 5+10 (13.9)	c h c (13.9)	g b b (5.7)
	N 7+9 2+12 (11.1)	c i a (11.1)	o b d (5.7)
	1 7+9 5+12 (5.6)	d b c (8.3)	b d b (5.7)
	N 6+8 5+10 (5.6)	c b a (2.8)	b d j (5.7)
	2* 7+9 5+10 (20.7)	c f c (20.7)	m b b (10.3)
	1 7+9 2+12 (10.3)	e b c (17.2)	m b j (6.9)
80	N 6+8 5+10 (10.3)	c h c (10.3)	b e a (6.9)
	N 7+9 5+10 (10.3)	c b c (6.9)	b g f (6.9)
	1 7 2+12 (6.9)	c f a (6.9)	l g f (6.9)
	1 7+8 5+10 (6.9)	c g a (6.9)	f l b (6.9)
	1 7+9 5+10 (6.9)	f j c (6.9)	b b a (3.45)
	2* 7+9 5+10 (30.4)	c b c (21.7)	b b b (13.0)
	N 7+9 5+10 (21.7)	c j c (13.0)	a e b (8.7)
90	1 7+9 5+10 (17.4)	c g a (8.7)	m l b (8.7)
	N 7+8 5+10 (13.0)	c g c (8.7)	a b b (4.3)
	1 7+9 2+12 (8.7)	c j a (8.7)	f b b (4.3)
	N 7+9 2+12 (8.7)	e j c (8.7)	b b g (4.3)
	2* 7+9 5+10 (25.6)	c b c (25.6)	b b b (16.3)
	2* 7+8 5+10 (14.0)	e b c (23.3)	m b b (13.95)
After 2000	2* 7+9 2+12 (11.6)	c j c (11.6)	m b j (9.3)
	1 7+8 5+10 (9.3)	c b a (7.0)	b l b (9.3)
	1 7+9 5+10 (9.3)	c f c (7.0)	g b b (4.7)
	N 7+9 5+10 (9.3)	d b c (4.7)	o b d (4.7)
	N 6+8 5+10 (4.7)	f b c (4.7)	b b h (4.7)

* - number in the parentheses indicates the percentage of frequencies

Table 2. Distribution of the most common spectra in decades

In order to improve the quality of new varieties, a narrowing of the genetic diversity was observed especially after the mass penetration of the variety Bezostaya 1 in the breeding programs after the 80's. The direct usage as a parent in the hybridization or indirectly as a participant in the pedigree of other varieties, led to unification of the spectra of new varieties with that of Bezostaya 1. It is necessary to broaden the genetic basis by including alleles with good influence over the wheat quality. Such alleles are *Glu-B1f* (subunit 13+16), *Glu-B1i* (subunit 17+18), *Glu-A3f*, *Glu-A3b*.

3. Progress in development of quality parameters of Bulgarian wheat varieties

The breeding for wheat quality started in 1962 when the first contemporary wheat breeding program in the Dobrudzha Agricultural Institute was adopted. In the two breeding centers (DAI and IPGR) 26 high quality varieties were created till now which was 1/3 of the entire variety list. During 2006 and 2008 sixty nine varieties were tested for their quality parameters. The varieties belonged to different quality groups according to Bulgarian State Standard (BSS). Based on the level of indices sedimentation value, wet glutenin content and valorimeter, varieties from this investigation were divided in three groups (table 3).

Level of indices	Varieties
	Strong wheat
sedim. > 47 WGC > 23 val. > 47	Aglika, Albena, Antitsa, Bezostaya 1, Goritsa, Ideal, Iveta, Laska, Lazarka, Merilin, Milena, Miziya, Pobeda, Slavyanka 196, Zlatina
	Medium wheat
sedim. 38 - 46 WGC 18 - 23 val. 40 - 47	Bolyarka, Charodeika, Enola, Karina, Liliya, Neda, Sadovo 552, Vratsa, Yanitsa, Yunak
	Weak wheat
sedim. < 38 WGC < 18 val. < 40	Antonovka, Kaliakra 2, Karat, Kristi, Petya, Pliska, Prelom, Svilena, Todora, Yantar, Zlatitsa

Table 3. Grouping of varieties according to the level of sedimentation (sedim.), wet glutenin content (WGC) and valorimeter (val.)

Some of the strong wheat varieties, according to BSS, failed to enter the first group in this investigation. Such varieties were: Demetra - with lower WGC; Preslav – with lower WGC and valorimeter; Dona – with lower sedimentation and WGC and Ludogorka and Momchil with lower levels for all three indices. A tendency in breeding of high quality varieties with low wet gluten content, even in comparison to medium and low-quality varieties, has been reported by Atanasova et al. (2010), Panayotov & Rachinsky (2002). One of the reasons for this is the use of Russian and Ukrainian sources in the breeding programs for wheat quality improvement. These sources have high quality of gluten but in lower amounts.

Wheat varieties, belonging to second group by BSS, in this investigation showed level of indices for the first group. With higher level of all three indices were varieties Bozhana and Stoyana. With higher sedimentation and valorimeter were Slaveya and Venka 1. Varieties Galateya, Prostor, Zagore had higher WGC and valorimeter value. Some varieties from third

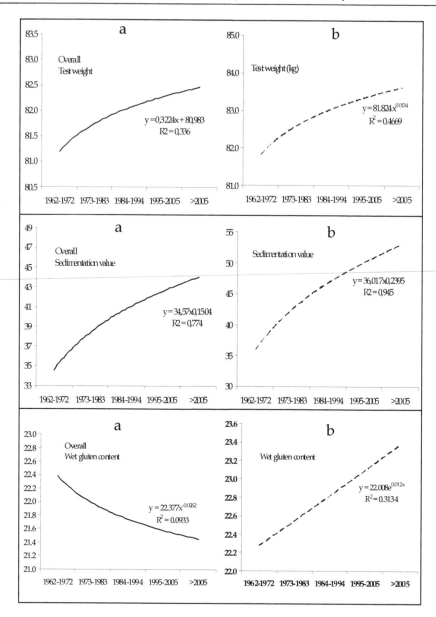

Fig. 2. Changes of quality indices by decades for the entire group (a) and for the quality group (b) of wheat varieties

group by BSS, also, showed higher level of some of the indices. For example varieties Trakiya and Ogosta had higher level of valorimeter value. Variety Pryaspa had higher WGC and varieties Charodeika and N 100-10 had higher level of the two indices. Similar discrepancies in the level of indices were observed in other investigations (Atanasova et al.,

2009a; Atanasova et al., 2010). This demonstrates the complex nature of the wheat quality, which is affected not only by genotype but by environments as well.

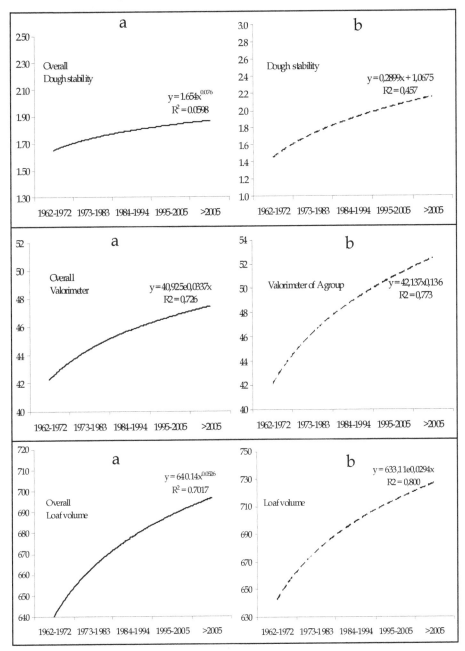

Fig. 2. Continue

As a result of breeding a progress was determined in almost all the indices defining end-use quality in wheat. The breeding periods were divided into decades as follows: 1962-1972; 1973-1983; 1984-1994; 1995-2005 and after 2005. The average values of the whole group of sixty-nine wheat varieties and that of the high-quality group were compared. The most strongly modification was observed for the indices: dough stability, where the enlargement was 24% for the entire group and 60% for the quality group; sedimentation value – 26%for whole group and 44% for quality group (figure 2). For the indices valorimeter and loaf volume a high level had been reached which changed very slowly through the years. The group of quality wheat varieties had greater progress compared to the all varieties regardless of the quality index and its behavior. It was valid even to the indices for which the breeding progress was not statistically proven.

4. Evaluation of genotype, environment and their interaction for quality parameters of Bulgarian varieties

The complex nature of wheat quality is determined by the level of various indices and dependent on a number of factors. One of the problems of breeding for quality is not only achieving a high level of the indices in the new varieties, but also achieving a stable expression of these indices in different years and environments (Atanasova et al., 2008; Johansson et al., 2001; Williams et al., 2008). It is valuable for breeding to determine the nature and direction of the effects of different genetic and environmental factors on the specific quality indices (Hristov et al., 2010; Panozzo & Eagles, 2000; Tsenov et al., 2004), and to study the performance of the varieties under changeable growing conditions. This would allow predicting to some degree their response to certain combinations of environmental factors (Atanasova et al., 2010; Drezner et al., 2006; Yong et al., 2004). The variation of the quality indices when growing each individual variety under different environments is always helpful both for the distribution of the new varieties and for their improvement in breeding programs (Gomez-Becera et al., 2010). The data on the variation of winter wheat quality is contradictable with regard to the level of genotype effect (Gomez-Beccera et al., 2010; Williams et al., 2008). In breeding there is always the question whether the high quality of a given variety under favorable environments is a prerequisite for the realization of its high-quality potential under unfavorable environments, too, or is it and impediment for its stability (Hristov et al., 2010; Tsenov et al., 2004).

Sixteen Bulgarian winter wheat varieties from first and second quality group according to BSS were tested at two locations: Dobrudzha Agricultural Institute – General Toshevo (DAI) and Institute of Agriculture and Seed Science, Obraztsov Chiflik, Rouse (OCH) during 2004–2007. The expression of 5 grain quality indices which give information about various quality aspects was analyzed: test weight (kg) (BSS 7971-2:2000), sedimentation value of flour (ml) (Pumpyanskii, 1971), wet gluten content in grain (%) (BSS 13375-88), valorimeric value (valorimeter, conditional units) (BSS 16759-88), and loaf volume, determined according to the methods adopted at the DAI laboratory.

The effect of the location was highest on wet gluten content and valorimeric value (table 4). The traits test weight and loaf volume were affected most by the year conditions. The distribution of the varieties into groups according to their quality had highest influence on the expression of sedimentation. The independent influence of the genotype on the variation of most of the investigated indices was lowest. Atanasova et al. (2008) and Zhang et al.

(2005) have found out that the genotype had higher effect on the expression of sedimentation, while Drezner et al. (2006, 2007), Panozzo and Eagles (2000), Tsenov et al. (2004) have proved that the environment has greater effect on the quality indices of wheat. Among the various combinations between the factors, the interaction *quality group x genotype* had the highest effect on almost all investigated traits. The exception was wet gluten content, which was affected most by the interaction *location x year*. Hristov et al. (2010), Mladenov et al. (2001), Williams et al. (2008) have found out that in spite of the significant effect of the interaction *genotype x environment* on the expression of the quality parameters of wheat, this effect was less significant than the independent influences of the genotype and the location.

Source of variation	TW	SDS	WGC	Val	Lvol
Main effects					
A: Location	161.1^{***}	1250.0^{***}	170.4^{***}	1058.0^{***}	43512.5^{***}
B: Year	205.5^{***}	876.6^{***}	91.6^{***}	173.0^{***}	72179.4^{***}
C: Group	76.3^{***}	2032.0^{***}	7.9^{ns}	666.1^{***}	39903.1^{***}
D: Genotype	11.2^{***}	28.9^{**}	22.2^{***}	48.7^{**}	3868.2^{***}
Interactions					
A x B	4.7^{***}	94.6^{***}	76.1^{***}	8.4^{ns}	703.6^{ns}
A x C	7.3^{***}	101.5^{***}	17.3^{**}	63.3^{**}	2032.0^{ns}
A x D	1.1^{*}	4.0^{ns}	1.5^{ns}	13.5^{ns}	942.4^{ns}
B x C	0.3^{ns}	15.6^{ns}	13.0^{**}	26.4^{*}	3650.5^{***}
B x D	0.4^{ns}	21.5^{**}	5.3^{ns}	9.0^{ns}	2273.9^{***}
C x D	17.4^{***}	172.7^{***}	13.2^{**}	157.4^{***}	11036.6^{***}

$*P<0.1; **P<0.05; ***P<0.01;$ *ns*-not significant

Table 4. Analysis of variance for mean squares of 16 wheat varieties

The growing conditions at DAI were more favorable for expressing genetic potential of wheat varieties as it is revealed by mean values of the indices (table 5). It is logical to expect higher values for the indices of the varieties from the first quality group. The results from the variance analysis were once again confirmed; the values for wet gluten content were not significant and there were no considerable differences between the two quality groups.

It is difficult in breeding for wheat quality to develop varieties with high quality indices, which remain stable under various growing conditions (Atanasova et al., 2008; Johansson et al., 2001; Williams et al., 2008). According to Finlay & Wilkinson (1963), Sudaric et al. (2006) genotypes with coefficient of regression (b_i) under 0.7 were considered unresponsive to changeable environments; with b_i between 0.7 and 1.3 had average stability; and >1.3 were considered responsive to good environments. Most of the varieties involved in this investigation have moderate stability for all the indices with some exceptions (table 6). There wasn't any regularity in manifestation of the stability. For some traits varieties showed stable reaction as b_i was near 1.0. For other traits varieties were specifically adapted to growing in unfavorable environments as b_i was below 0.7. For example variety Enola was with stable reaction for test weight, sedimentation value, WGC (b_i was 1.01, 1.15, 1.08, respectively) and with b_i=0.598 for loaf volume.

Factors	TW	SDS	WGC	Val	Lvol
Location					
DAI	81,2[b]	39,4[b]	21,6[b]	43,7[b]	673,8[b]
OCH	78,9[a]	33,1[a]	19,3[a]	38[a]	636,9[a]
Year					
2004	81,5[c]	36,5[b]	20,0[a]	42,4[b]	641,1[a]
2005	76,5[a]	43,4[c]	22,4[b]	41,8[b]	725,6[b]
2006	80,2[b]	31,3[a]	18,4[a]	41,7[b]	633,1[a]
2007	82,2[c]	33,8[ab]	21,0[ab]	37,4[a]	621,7[a]
Group of quality					
I-group	80,8[b]	40,2[b]	20,7[a]	43,1[b]	673,0[b]
II-group	79,3[a]	32,3[a]	20,2[a]	38,5[a]	637,7[a]

Numbers with different letters differ significantly

Table 5. Mean values of the investigated traits according to different factors of variation

The valorimeric value is an index which determines gluten quality and which correlates strongly with end-use quality. Most of the investigated varieties possess moderate stability (b_i from 0.7 to 1.3) with regard to this trait. Varieties Milena and Aglika (b_i = 1.00 and b_i = 1.25, respectively) from first quality group were an evidence that high values of the traits can be observed in stable genotypes, too (Hristov et al., 2010; Mladenov et al., 2001; Sudaric et al., 2006) and that stability is not necessarily related to low mean values, as previously stated (Becker and Leon, 1988). The ecovalence values of the individual varieties for this trait were very high, which brings forth the strong effect of the environment. The variation expressed through the ecovalence was much lower in the wheat varieties from the second group in comparison to the varieties from the first group (56.8 and 99.6, respectively) (table 6).

A number of researchers have found out that the varieties with lower quality potential are considerably more stable, especially under stress (Atanasova et al., 2010; Tsenov et al., 2004). The values of the regression coefficient of the separate groups are quite indicative in this respect. The regression coefficient in the first group was above 1.0. This means that that the varieties from this group realized their quality potential under conditions more favorable for its formation. Therefore the W_i values of variety Zlatina, which is a quality variety, were several times higher than the mean values for this group. On the other hand, varieties like Aglika, Milena and Preslav demonstrated considerably lower variation of their quality indices in comparison to the other varieties from the group. It can be suggested that such stability is due to the fact that the absolute values of the indices in this case were low. These data once again prove that when comparing the response of the varieties to the environment with regard to their quality, conclusions should be made very carefully. The results for the individual varieties and their indices are not unidirectional and therefore when selecting a variety for mass production it is necessary to consider the purpose for its cultivation, the potential of the given variety and its ability to realize this potential with stability and adequacy.

Variety	Test weight		Sedimentation		Wet gluten content		Valorimetric value		Loaf volume	
	bi	Wi	bi	Wi	bi	Wi	bi	Wi	bi	Wi
Igroup										
Pobeda	1.09	2.75	0.84	8.48	1.174	21.2	1.552	121.2	1.300	7.04
Albena	0.92	1.85	1.29	15.4	0.854	12.2	1.020	59.67	0.650	9.67
Preslav	1.05	2.73	1.43	8.01	1.226	9.37	0.648	26.26	1.015	10.59
Milena	1.07	1.54	0.91	10.7	1.219	36.5	1.000	21.33	0.834	4.56
Aglika	1.26	6.24	1.01	18.2	1.022	10.5	1.251	89.42	1.166	12.45
Progres	1.40	9.48	1.28	6.39	0.656	36.7	0.700	42.86	1.283	9.32
Zlatina	1.10	2.40	1.11	10.2	0.800	23.1	1.855	274.2	1.350	12.36
Demetra	0.95	3.16	1.62	22.3	1.709	40.4	0.754	162	0.987	2.46
Average	1.104	3.77	1.186	12.46	1.080	23.72	1.100	99.62	1.073	8.56
II group										
Sadovo 1	1.3	5.24	1.21	5.52	1.154	14.5	1.331	133.1	1.303	5.45
Enola	1.01	3.53	1.15	16.3	1.078	82.6	0.963	86.11	0.598	11.00
Pliska	1.13	2.24	1.20	11.5	0.645	47.0	0.814	14.58	1.019	4.22
Pryaspa	1.07	3.00	0.88	5.33	1.229	14.3	0.912	23.51	0.905	10.67
Yantar	0.4	25.8	0.65	6.46	0.860	15.4	0.658	41.83	0.909	2.72
Kristi	0.61	8.35	0.75	5.86	0.900	25.6	0.736	70.67	0.789	8.09
Prelom	0.85	3.05	0.85	9.18	0.671	19.3	0.648	28.39	0.960	11.77
Boryana	1.19	3.52	0.69	2.88	0.803	86.6	1.158	56.17	1.198	8.69
Average	0.944	6.84	0.923	7.88	0.920	38.16	0.900	56.79	0.960	7.83

Table 6. Stability parameters of varieties from different quality groups

When compare the mean values of varieties with their stability it can be seen that for the conditions of DAI varieties Demetra, Zlatina, Aglika differ from the total group with higher levels of the indices (figure 3). For the other location (OCH) these varieties were Demetra, Enola, Aglika. The most poorly performing variety for the two locations was Prelom.

Projection to the ordinate of the two-dimensional matrix genotype x character, regardless of the direction, determines the stability of the varieties (Yan et al., 2007). For the conditions of DAI as most stable and exceeding the mean level were varieties Aglika, Sadovo 1, Preslav, Demetra. Varieties Milena and Preslav had most stable reaction at OCH location as they had almost null projection to the ordinate. Under the same conditions with high variability stand out varieties Enola and Pliska.

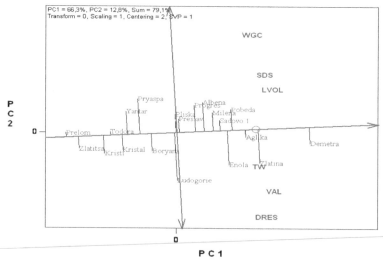

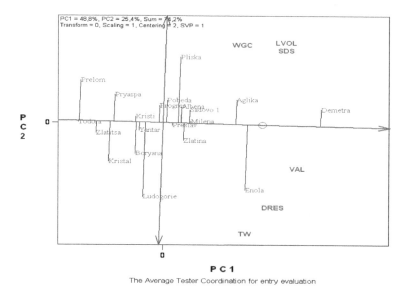

Fig. 3. Comparison between mean values of the varieties and their stability for investigated indices for conditions of DAI (up) and OCH (down)

Another investigation was carried out during 2001-2003 with 10 varieties with different quality in two typical for grain production in Bulgaria regions – Dobrudzha Agricultural Institute (DAI) in North-East Bulgaria and Chirpan in South Bulgaria. Data presentation by Principal Component Analysis (PCA) allows more comprehensive explanation of the variation of quality indices of studied varieties in the two locations (Stoeva et al., 2006). There was a similar component structure of the factors which mainly influence the variation of the quality over locations. About 65% of the total variation of the two-dimensional matrix variety x character for each location is for first and second PC. The biggest part is for PC1 (table 7).

Indices	DAI		Chirpan	
	PC1	PC2	PC1	PC2
Test weight, kg	0.647	-0.385	0.587	0.395
Vitreousness, %	0.603	-0.459	0.358	0.845
Endosperm hardiness, %	0.835*	-0.258	0.674	0.678
Wet gluten content, %	0.493	0.421	0.325	-0.361
Sedimentation value, ml	0.910**	0.206	0.916**	-2,362
Dough stability, min	0.805*	1,841	0.971**	-2,149
Valorimeric value,	0.922**	2,983	0.990**	-2,017
Bread loaf, ml	0.778*	0.181	0.742*	-0.505

Table 7. Correlation between quality indices, PC1 and PC2 over locations

Indices endosperm hardiness, sedimentation value, dough stability, valorimeter and bread loaf were determinative for the quality of the studied varieties in DAI as the levels of the PC1 were positive and higher than 0.700. For the other location (Chirpan) these were sedimentation value, dough stability, valorimeter and bread loaf.

Varieties from this investigation had different reaction to the environment in the two locations. Varieties with lower quality such as Pryaspa, Karat, Todora showed low variability connected with not linear reaction to the environments as their PC1 and PC2 were negative (table 8). High quality varieties Aglika, MIlena, Galateya showed high stability in the two locations irrespective of the direction in the changing environments. The different sing in the reaction of varieties show that some of them react specifically to the environments. For example in poor environments some varieties managed to save higher levels of indices and vice versa.

Variety	DAI		Chirpan	
	PC1	PC2	PC1	PC2
Yantar	0.334	0.684	0.452	0.731
Pryaspa	-0.166	-0.490	-0.121	-0.395
Milena	0.722	0.764	0.801	0.884
Aglika	2.001	-0.103	1.120	0.030
Galateya	0.922	-0.386	0.793	-0.268
Albena	-0.507	0.957	-0.317	0.784
Enola	-0.080	1.130	0.009	0.922
Karat	-0.406	-0.814	-0.566	-0.753
Kristal	-1.668	0.388	-1.284	0.299
Todora	-0.493	-2.131	-0.569	-1.977

Table 8. Values of principal components of varieties

Wheat growing depends on environments although the right technologies and new varieties can manage to lower this dependence. That's why it is valuable to select varieties which give high quality in different environments.

5. Breeding of wheat quality in Bulgaria – Steps, drawbacks, prospects

Breeding of new varieties with good quality has always been on the attention of the breeding programs in Bulgaria (Boyadjieva et al., 1999; Panayotov et al., 1994; Rachinski, 1966; Stoeva & Ivanova, 2009). Combining high yield and quality in wheat is a challenge for the contemporary breeding as it is connected with many obstacles with different nature (Baenziger et al., 2001; Dencic et al., 2007; Eagles et al., 2002; Trethowan et al., 2001). Analysis of wheat quality begins with studying the initial material (Tsenov et al., 2010c). When foreign samples are received they are study for three years for a set of quality indices. Those with high levels are included in hybridization programs especially if they show high yield potential, stress tolerance, disease resistance. The basic foreign parent components for hybridization are from breeding centers with traditions and excellent achievements in breeding for wheat quality (table 9).

The quality of Bulgarian wheat varieties is based on the widespread use of variety Bezostaya 1 and other Russian and Ukraine sources (Panayotov & Kostov, 2007; Todorov et al., 1998; Tsenov et al., 2010b). During the last years samples with different origin are widely used as they possess high level of productivity combined with high end-used quality. The quality analysis during the breeding process is made according the scheme described in the table 10.

The beginning of quality analysis of the breeding lines is in the screening nursery. The main tests in this unit are sedimentation value of the flour and grain protein content. Quality index of each line and quality standard varieties is calculated by these two parameters. These indices are defined as they possess high and positive correlation with the main parameters, connected with the strength of flour and dough (Bona et al., 2003, Dacheva & Boydjieva, 2002).

Country	Grain properties	Dough properties	Gluten properties	Bread making	End-use quality
Bulgaria	*	*	*	*	*
Romania	*		*		*
Serbia	*				*
Turkey	*		*		*
Odessa, Ukraine	*	*	*	*	*
Harkov, Ukraine	*	*		*	
Mironovka, Ukraine		*		*	*
Ktrasnodar, Russia	*	*	*		*
Nebraska, USA	*	*		*	
Oklahoma, USA	*	*		*	
Texas, USA	*	*		*	
Australia		*		*	
Canada		*		*	
Argentina	*		*		*

Table 9. Origin of the initial breeding material for wheat quality

Quality traits	Screening Nursery (SN)	Preliminary Yield Trail (PYT)	Competitive Yield Trails (CYT)
Sedimentation value	+	+	+
Grain Protein Content	+	+	+
Quality index*	+	+	+
Test weight		+	+
Farinograph characteristics			+
End-use quality			+
Bread making characteristics			+
SDS-PAGE			+

* Quality index=Sedimentation/Protein content

Table 10. Quality indices, used for breeding in the separate trial units

In the next trial units (PYT, CYT) the analysis extend and cover almost all aspects of quality of grain, flour, dough and bread. Along with this an analysis is made to determine the allelic diversity of the lines. At least three year data from CYT, which is the last level of screening of breeding materials, are needed in order to be able to assign each line in the following quality group: A – strong wheat; B – medium and C – wheat with soft endosperm (figure 4).

In the yield trails of Executive Agency of Variety Testing, Field Inspection and Seed Control (EAVTFIS) the candidate-varieties are also testing in these three groups at least 2 or 3 years. Each group has specific standards for comparison.

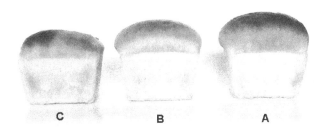

Fig. 4. Bread-making quality groups

The evaluation and selection for quality is associated with many difficulties from different nature. They can be determined as follows:

1. There is a wide genetic variation for each index caused by three genomic structure of the crop.
2. Environments have an enormous share in wheat quality formation and embarrassed the genetic expression of the varieties (Dencic et al., 2011, Tayyar, 2010).
3. Wheat quality is reduced in the presence of biotic and abiotic stress (Atanasova et al., 2010).
4. The inheritance of quality indices is complicated and polygenic (Tsenov, 1994, Tsenov et al., 1995, Tsenov & Stoeva, 1997).

5.   Samples with high combining ability for quality are limited.
6.   Complex genetics suggests a special selection procedure (Tsenov & Stoeva, 1998).
7.   The selection in the early hybrid populations is not effective because of fading forming process and restricted amount of grain for analysis.
8.   There is a lack of breeding indices for parallel selection for quality and yield potential and other parameters.

6. Conclusions

As a result of a nearly half a century breeding in Dobrudzha Agricultural Institute 26 varieties from the first quality group are created. This is nearly one third of the entire variety list. The massive usage of variety Bezostaya 1 in the breeding programs direct or indirect narrowed the genetic diversity of the materials and to some extent unified their glutenin and gliadin spectra. The efforts to include foreign initial materials in hybridization, with alleles with good influence over the wheat quality, will probably be rewarded in the near future.

A breeding progress is determined for almost all the indices, defining end-use quality in wheat. The group of high quality wheat varieties has greater progress compared to all varieties irrespective of the indices.

Different factors influence on the wheat quality and on the expression of different indices. The knowledge of the nature and direction of the effect of these factors is valuable for predicting to some extend the performance of wheat varieties in certain growing conditions. Most of the Bulgarian varieties show moderate stability at different environments. But the results for the individual varieties and their indices are not unidirectional and therefore when selecting a variety for mass production it is necessary to consider the purpose for its cultivation, the potential of the given variety and its ability to realize this potential with stability and adequacy.

7. Acknowledgment

The authors give their acknowledgments to all colleagues from the different laboratories for their help in determine different aspects of wheat quality.

8. References

Atanasova, D., Tsenov, N., Stoeva, I. & Dochev, V. 2008. Genotype x environment interaction for some quality traits of Bulgarian winter wheat varieties, In: *Modern Variety Breeding for Present and Future Needs,* J. Prohens and M. L. Badenes (Eds), Proceedings of *The 18th EUCARPIA General Congress,* 9-12 September 2008, Valencia, Spain, pp. 532-537.

Atanasova, D., Dochev, V., Tsenov, N. & Todorov, I. 2009a. Influence of genotype and environments on quality of winter wheat varieties in Northern Bulgaria. *Agricultural science and technology,* 1 (4): 121-125.

Atanasova, D., Tsenov, N., Todorov, I. & Ivanova, I. 2009b. Glutenin composition of winter wheat varieties bred in Dobrudzha Agricultural Institute. *Bulgarian Journal of Agricultural Science,* 15 (1): 9-19.

Atanasova, D., Tsenov, N., Stoeva, I. & Todorov, I. 2010. Performance of Bulgarian winter wheat varieties for main end-use quality parameters under different environments. *Bulgarian Journal of Agricultural Science*, 16 (1): 22-29.

Baenziger, P.S., Shelton, D.R., Shipman, M.J. & Graybosch, R.A. 2001. Breeding for end-use quality: Reflections of the Nebraska experience. *Euphytica*, 119 (1-2): 95-100.

Becker, H.C. & Leon, J. 1988. Stability analysis in plant breeding. *Plant Breeding*, 101: 1-23.

Bona, L., Matuz, J. & Acs, E. 2003. Correlations between screening methods and technological quality characteristics in bread wheat. *Cereal Research Communications*, 31 (1-2): 201-204.

Boyadjieva, D., Dacheva, V. & Mungova, M. 1999. A study on the rheological properties of T. aestivum germplasm from different geographic areas. Bulgarian Journal of Agricultural Science, 5: 703-712.

Bradova, J. 2008. Allelic diversity of HMW- and LMW-glutenin subunits wheat varieites (Triticum aestivum L.) registered in the Czech Republic. In: *Modern Variety Breeding for Present and Future Needs*, J. Prohens and M. L. Badenes (Eds), *Proceedings of The 18th EUCARPIA General Congress*, 9-12 September 2008, Valencia, Spain, pp. 553-557.

BSS 7971-2:2000 – Bulgarian State Standard for test weight.

BSS 13375-88 - Bulgarian State Standard for wet gluten content in grain.

BSS 16759-88 - Bulgarian State Standard for valorimeric value.

Bushehri, A.A.S., Gomarian, M.& Samadi, B. Y. 2006. The high molecular weight glutenin subunit composition in old and modern bread wheats cultivated in Iran. *Australian Journal of Agricultural Research*, 57: 1109-1114.

Dacheva, V. & Boydjieva, D. 2002. The quality index as a selection tool in winter wheat (T. aestivum L.) breeding. *Proceedings of "120 years Agricultural Science in Sadovo", Scientific Reports*, pp. 88-91. (In Bul).

Dencic, S., Kobiljsli, B., Mladenov, N., Hristov, N. & Pavlovic, M. 2007. Long-term breeding for bread making quality in wheat. In: *Wheat production in stressed environments*, Buck, H.T., Nisi, J.E., Salomyn, N. (Eds.), pp: 495-501.

Dencic, S. Mladenov, N. & Kobiljski, B. 2011. Effects of genotype and environment on bread making quality in wheat. *International Journal of Plant Production*, 5 (1): 71-81.

Drezner, G., Dvojkovic, K., Horvat, D., Novoselovic, D., Lalic, A., Babic, D. & Kovacevic, J. 2006. Grain yield and quality of winter wheat genotypes in different environments. *Cereal Research Communications*, 34 (1): 457-460.

Drezner, G., Dvojkovic, K., Horvat, D., Novoselovic, D. & Lalic, A. 2007. Environmental impacts on wheat agronomic and quality traits. *Cereal Research Communications*, 35 (2): 357-360.

Eagles, H.A., Hollamby, G.J., Eastwood, R.F. 2002. Genetic and encironmental variation for grain quality traits routinely evaluated in southern Australian wheat breeding programs. *Australian Journal of Agricultural Research*, 53: 1047-1057.

Finlay, K. W. & Wilkinson, G. 1963. The analysis of adaptation in a plant breeding programme. *Australian Journal of Agricultural Research*, 14, 742-754.

Gomez-Beccera, H.F., Abudalieva, A., Morgunov, A., Abdullaev, K., Bekenova, L., Yessimbekova, M., Sereda, G., Shpigun, S., Tsygankov, V., Zelenskiy, Y., Pena, R.J. & Chakmak, I. 2010. Phenotypic correlations, G x E interactions and broad sense heritability analysis of grain and flour quality characteristics in high latitude spring bread wheats from Kazahstan and Siberia. *Euphytica*, 171 (1): 23-38.

Hristov, N., Mladenov, N., Djuric, V., Kondic-Spika ,A., Marjanovic-Jeromela, A. & Simic, D. 2010. Genotype by environment interactions in wheat quality breeding programs in southeast Europe. *Euphytica*, 174 (3): 315-324.

Jackson, E.A., Holt, L.M. & Payne, P.I. 1983. Characterization of high-molecular-weight gliadin and low-molecular-weight glutenin subunits of wheat endosperm by two-dimensional electrophoresis and the chromosomal localization of their controlling genes. *Theoretical and Applied Genetics*, 66: 29-37.

Johansson, E., Prieto-Linde, M.L. & Jonsson, J.O. 2001. Breeding for stability in bread milling quality. In: *Wheat in a Global Environment*, Z. Bedo and L. Lang (eds.) Kluwer Academic Publishers, Netherlands, 229-235.

Knoblochova, H. & Galova, Z. 2000. High molecular weight glutenin variation and relative quantization in winter spelt wheat cultivars. *Rostlinna Vyroba*, 46: 255-260.

Lawrence, G.L. & Shepherd, K.W. 1981. Inheritance of glutenin protein subunits of wheat. *Theoretical and Applied Genetics*, 60: 333-337.

Li, L., He, Z., Yan, J., Zhang, Y, Xia, X & Pena, R.J. 2005. Allelic variation at the Glu-1 and Glu-3 loci, presence of the 1B.1R translocation, and their effects on mixographic properties in Chinese bread wheats. *Euphytica*, 142(3): 197–204.

Masci, S., Rovelli, L., Kasarda, D.D., Vensel, W.H. & Lafiandra, D. 2002. Characterisation and chromosomal localisation of C-type low-molecular-weight glutenin subunits in the bread wheat cultivar Chinese Spring. *Theoretical and Applied Genetics*, 104: 422-428.

Mladenov, N., Misic, T., Przulj, N. & Hristov, N. 2001. Bread-making quality and stability of winter wheat grown in semiarid conditions. *Rostlinna vyroba*, 47: 160-166.

Panayotov, I., Boyadjieva, D., Todorov, I., Stankov, I., Dechev, D. & Tsvetkov, S. 1994. Situation and problems in wheat breeding in Bulgaria, *Plant Science*, 31 (3-4): 48-57.

Panayotov, I. & Kostov, K. 2007. Combining Bulgarian and Ukrainian breeding for improving wheat quality and productivity. *Proceedings of International Scientific Conference "Plant Genetic Stocks – The Basis of Agriculture of Today"*, 13-14 June 2007, Sadovo, Bulgaria, vol. 2-3: 371-374.

Panayotov, I. & Rachinsky, T. 2002. Wheat breeding as a basis of grain production in Bulgaria. *Proceedings of 50th Anniversary of Dobrudzha Agricultural Institute, "Breeding and Agrotechnics of field crops"*, Dobrich, June 2001, vol. 1: 21-37. (Bg)

Panayotov, I., Todorov, I., Stoeva, I. & I. Ivanova. 2004. High quality wheat cultivars created in Bulgaria during the period 1994-2004 - achievements and perspectives. *Field Crops Studies*, 1 (1): 13-19 (In Bul).

Panozzo, J.F. & Eagles, H.A. 2000. Cultivar and environmental effects on quality characters in wheat. 2. Protein. *Australian Journal of Agricultural Research*, 51, 629–636.

Pumpyanskii, A.Y. 1971. Technological characteristics of common wheat. L., pp. 22 (In Rus).

Rachinski, T. 1966. Results from testing of some Russian wheat varieties in North-West Bulgaria. *Plant Science*, 3 (10): 11-21. (In Bul)

Singh, N.K. & Shepherd, K.W. 1988. Linkage mapping of the genes controlling endosperm proteins in wheat. 1. Genes on the short arms of group 1 chromosomes. *Theoretical and Applied Genetics*, 66: 628-641.

Stoeva, I. & Ivanova, A. 2009. Interaction of the technological properties of common winter wheat varieties with some agronomy factors. *Bulgarian Journal of Agricultural Science*, 15 (5): 417-422.

Stoeva, I., Tsenov, N., & Penchev, E. 2006. Environmental impact on the quality of bread wheat varieties. *Field Crop Studies*, 3 (1): 7-17.

Sudaric, A., Simic, D. & Vrataric, M. 2006. Characterization of genotype by environment interactions in soybean breeding programmes of southeast Europe. *Plant Breeding*, 125, 191-194.

Tabasum, A., Iqbal, N., Hameed, A. & Arshad, R. 2011. Evaluation of Pakistani wheat germplasm for bread quality based on allelic variation in HMW glutenin subunits. *Pakistan Journal of Botany*, 43 (3): 1735-1740.

Tayyar, S. 2010. Variation in grain yield and quality of Romanian bread wheat varieties compared to local varieties in north-western Turkey. *Romanian Biotechnological Letters*, 15(2): 5189-5196.

Todorov I. 2006. Investigation of grain storage proteins and their use as markers in wheat breeding. *PhD Thesis*, DAI, General Toshevo, pp. 398 (In Bul).

Todorov, I., Ivanov, P. & Ivanova, I. 1998. The HMW-GS composition of wheat genotypes from Western Europe. *Proceedings of 2nd Balkan Symposium on Field crops*, Novi Sad, 16-20 June, vol. 1: 33-37.

Todorov, I., Ivanov, P.& Ivanova, I. 2006. Genetic diversity of high molecular weight glutenin alleles in varieties with different origin, *Field Crop Studies*, III (4): 487-498 (In Bul).

Tohver M. 2007. High molecular weight (HMW) glutenin subunit composition of Nordic and Middle European wheats. *Genetic Resources and Crop Evolution*, 54: 67-81.

Trethowan, R., Pena, R.J. & van Ginkel, M. 2001. The effect of indirect tests for grain quality on the grain yield and industrial quality of bread wheat. *Plant Breeding*, 120 (6): 509-512.

Tsenov, N. 1994. Combining ability of the main economic traits associated with productivity and quality in combinations of winter wheat varieties. *PhD Thesis*, pp. 185 (In Bul).

Tsenov, N., Atanasova, D., Todorov, I., Ivanova, I. & Stoeva, I. 2009. Allelic diversity in Bulgarian winter wheat varieties based on polymorphism of glutenin subunit composition, *Cereal Research Communications*, 37 (4): 551-558.

Tsenov, N., Atanasova, D., Todorov, I., Ivanova, I. & Stoeva, I. 2010a. Quality of winter common wheat advanced lines depending on allelic variation of Glu-A3. *Cereal Research Communications*, 38 (2): 250-258.

Tsenov, N., Atanasova, D., Stoeva, I. & Petrova, T. 2010b. Grain yield, end-use quality and stress resistance of winter wheat cultivars Aglika and Slaveya. *Scientific Works Agricultural University*, Plovdiv, 55 (1): 27-34.

Tsenov, N., Stoeva, I. & Atanasova, D. 2010c. Breeding of end-use quality of winter wheat in Dobrudzha Agricultural Research Institute – present and prospects, *Field Crop Studies*, 6 (2): 217-233 (In Bul).

Tsenov, N., Kostov, K., Gubatov, T. & Peeva, V. 2004. Study on the genotype x environment interaction in winter wheat varieties. I. Grain quality. *Field Crop Studies*, 1 (1), 20-29. (In Bul).

Tsenov, N. & Stoeva, I. 1997. Potentialities to combine high productivity with grain quality of winter common wheat hybrids. *Plant Science*, 34 (2): 9-13 (In Bul).

Tsenov, N. & Stoeva, I. 1998. Response to selection for improvement of grain quality in winter bread wheat crosses. *Proceedings of the 9th International Wheat Genetics Symposium*, Saskatoon, Canada, 2-7 August, Vol. 4, pp. 285-287.

Tsenov, N., Stoeva, I. & Penchev, E. 1995. Combining ability for sedimentation value in diallel crosses of winter bread wheat (Triticum aestivum). *Proceedings of "The first Balkan symposium on breeding and cultivation of wheat, sunflower and legume crops"*, 26 - 29. 06, Albena, Bulgaria: 244-247.

Williams, R. M., O'Brien, L., Eagles, H.A., Solah, V.A. & Jayasena, V. 2008. The influence of genotype, environment, and genotype x environment interaction on wheat quality. *Australian Journal of Agricultural Research*, 59 (2), 95-111.

Yan, W., M. S. Kang, B. Ma, S. Woods and P.L. Cornelius. 2007. GGEbiplot vs. AMMI analysis of genotype-by-environment data. Crop Science, 47: 643-655.

Yong, Z., Zhonghu, H., Ye, G., Aimin, Z. & van Ginkel, M. 2004. Effect of environment and genotype on bread-making quality of spring-sown spring wheat cultivars in China. *Euphytica*, 139 (1): 75-83.

Zhang Y, Zhang Y, He Z and Ye G, 2005. Milling quality and properties of autumn-sown Chinese wheats evaluated through multi-location trials. *Euphytica*, 143 (1-2), 209-222.

2

Hevea Germplasm in Vietnam: Conservation, Characterization, Evaluation and Utilization

Lai Van Lam, Tran Thanh, Le Thi Thuy Trang,
Vu Van Truong, Huynh Bao Lam and Le Mau Tuy
Rubber Research Institute of Vietnam
Ho Chi Minh City
Vietnam

1. Introduction

Germplasm collections have provided original materials for plant breeding program and crop improvement. Because of their genetic diversity and possible occurrence of particular desirable genes, germplasm collections are useful targets for plant breeders as well as other biologists. Recently, many of germplasm are being lost worldwide due to habitat destruction, invasion of foreign species, and reliance on fewer high yielding strains. Therefore, maintaining germplasm of agricultural crops is very important. Normally, a germplasm collection will be utilized based on its characters of immediate perceived value or its potential variation. Furthermore, it can also be used to better understand the properties and performances of the plants, particularly at the genomic level. Nowadays, countries all over the world have set up facilities for conservation, characterization and utilization of germplasm collections of various crops either directly or indirectly.

Rubber tree (*Hevea brasiliensis* Willd. ex A. de Juss. Müell. Arg) which produces natural rubber is a tall deciduous perennial tree belonging to the Euphorbiaceae family. *Hevea brasiliensis* is a native of the Amazon basin and was introduced to countries in the tropical belts of Asia and Africa during late 19th century. It can be termed as one of the most far reaching and successful introductions in plant history, resulting in plantations with about 10.6 million hectares in the world for providing the industry with natural rubber of 9.62 million tonnes in the year 2009 (IRSG, 2009). Approximately 78% of the cultivation has taken place in Southeast Asia, 15% in other Asia countries, 5% in Africa, and 2% in Latin America (IRSG, 2009). The main producing countries were Thailand, Indonesia, Malaysia, India and Vietnam (IRSG, 2010). As the fifth natural rubber producer in the world, Vietnam produced 723,700 tonnes that shared about 7.5% of the world's natural rubber production and total area under rubber trees were 674,200 hectares that shared 6.1% of the world's rubber area (Hoa, 2010). In Vietnam, areas under rubber trees are mainly in the South-eastern region (65.2%), followed by the Highlands (23.4%), central coastal area (9.7%) and the new areas developed in the North-western region (1.6%) (Hoa, 2010).

Hevea brasiliensis was introduced into Vietnam in 1897 from Bogor (Indonesia) by Alexandre Yersin. Since then, rubber tree has been considered as one of the most important crops and widely cultivated throughout the country, particularly in the Southeast and Highlands regions of Vietnam. Currently, about 3,500 accessions of *Hevea brasiliensis* have been collected and conserved in Vietnam. The majority of this germplasm were derived from the IRRDB'81 germplasm collected in the Amazonian habitats of the genus. This collection has been considered as the key factor contributing to the improvement of rubber tree through breeding programs. The characterization and evaluation of the germplasm are considered to be important aspects of *Hevea* germplasm conservation. Without proper characterization and evaluation, valuable genetic variation in the collections cannot be used for rubber improvement effectively. Understanding the genetic diversity of different genetic resources of *H. brasiliensis* would be important in `order to optimize their management and to utilize *Hevea* germplasm in breeding programs.

2. *Hevea* genetic resources

The genus *Hevea* is basically composed of 10 species: *H. brasiliensis, H. guianensis, H. benthamiana, H. pauciflora, H. spruceana, H. microphylla, H. rigidifolia, H. nitida, H. camporum,* and *H. camargoana* (Schultes, 1990). According to Clement-Demange et al. (2000), it is generally considered that there is no biological barrier between them, and some species proved to be inter-crossable by hand-pollination; therefore, the *Hevea* species can be considered as a species complex. *H. paludosa* has been identified in Brazil by Ule in 1905 as the eleventh species (Gonçalves et al., 1990; Priyadarshan and Gonçaalves, 2002). A review on the elaborate description of taxonomical and botanical aspects of *Hevea* has been reported by Schultes (1977, 1987, 1990) and Wycherley (1992). As its natural habitat, *Hevea* species have presented in Brazil, Bolivia, Peru, Ecuador, Colombia, Venezuela, Surinam and French Guiana as shown in Figure 1. These *Hevea* species have 2n = 36 chromosomes, with the possible exception of one triploid clone of *H. guianensis* (2n = 54) and the possible existence of one genotype of *H. pauciflora* with 2n = 18 (Baldwin, 1947; Majumder, 1964), and *H. brasiliensis* behaves as an amphidiploid (Ramaer, 1935; Ong, 1975; Wycherley, 1976).

It is known that all high-yielding cultivars of rubber tree (*Hevea brasiliensis* Muell. Arg.) in the world originated from breeding programs initially developed in Southeast Asia with a very narrow genetic base. Historically, the introduction of the rubber trees into Asia began with the transfer of 70,000 seeds to England by Henry Wickham in 1876 (Wycherly, 1968). During the years of 1876-1877, a total of 2,397 *Hevea* seedlings were sent to several Asian countries such as Sri Lanka (1,919 seedlings), Bogor (Java) (18 seedlings) and Singapore (22 seedlings) (Wycherly, 1968; Dean, 1987; Baulkwill, 1989). Because of the centrally geographical position of Singapore in Asia and the influences of the British to the development of the rubber industry during these years, the collection of 22 seedlings planted in the Singaporean botanical garden became the main source of the rubber trees which was distributed to and planted in Asian countries later. Since this introduction, the rubber tree has become an important perennial crop as the major source of commercial rubber in the world (Fig. 2) and this collection was mentioned as Wickham collection (W). Since the current *Hevea* varieties all came from such a single population, it is necessary to enlarge the genetic basis for *Hevea* breeding program.

In order to enlarge the genetic basis for *Hevea*, a large collection of *H. brasiliensis* accessions from various areas in Colombia was gathered by Schultes after 1945 and then duplicated in

Ivory Coast (Nicolas, 1985). In addition, a collecting survey in the Madre de Dios basin in Peru was organized in 1948 by the Peruvian Ministry of Agriculture (Rands and Polhamus, 1955); the resulting seedlings were multiplied by grafting, and then introduced and studied in Liberia, Guatemala and Brazil under the name of MDF accessions (Bos and McIndoe, 1965). Moreover, another collecting expedition was carried out in the Brazilian states of Acre and Rondonia by both the Brazilian Agricultural Research Corporation (EMBRAPA) and French Institute for Rubber Research (IRCA) in 1974 (Hallé and Combe, 1974; Seguin et al., 2003); 42 resulting wild elite-tapped trees were collected and propagated by grafting under the name as EMBRAPA/IRCA accessions. Remarkably, in 1981, the International Rubber Research and Development Board (IRRDB) conducted an expedition covering three western states of Brazil, namely Acre (AC), Rondonia (RO), and Mato Grosso (MT), in 16 different districts and in 60 different locations overall to collect wild *Hevea* germplasm (refered as IRRDB'81 collection). As a result, a total of 63,768 seeds, 1,413 meters of budwood from 194 high yielding trees and 1,160 seedlings were collected (Tan, 1987; Simmonds, 1989; Onokpise, 2004). Of these, 12.5% and 37.5% of the seeds were sent to Malaysia and Ivory Coast, respectively, and the remaining 50% of the collections were maintained in Brazil (Clément-Demange et al., 2007). The genotypes issued from budwood collection were also then brought to Malaysia and Ivory Coast. The collection planted in Malaysia and Ivory Coast has then been distributed to all IRRDB members as clones since 1984.

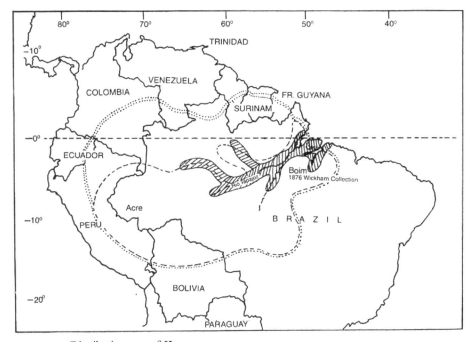

.... Distribution area of *Hevea*

- - - - - Distribution area of *Hevea brasiliensis*

Fig. 1. Geographical origin of *Hevea* (adapted from George (2000))

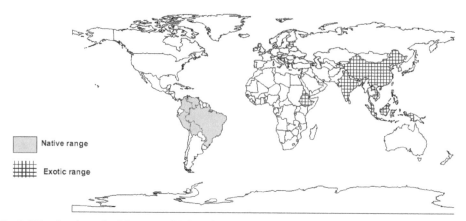

Fig. 2. Distribution of rubber tree (*H. brasiliensis*) in the world where this species has been planted (adapted from Orwa et al. (2009)). Native range: Bolivia, Brazil, Colombia, Peru and Venezuela; Exotic range: Brunei, Cambodia, China, Ethiopia, India, Indonesia, Laos, Liberia, Malaysia, Myanmar, Philippines, Singapore, Sri Lanka, Thailand, Uganda, Vietnam, Guatemala, Cameroon, Ivory Coast, Ghana, Gabon, Guinea, Liberia, Nigeria, Congo, Bangladesh, Papua New Guinea, and Mexico.

3. Long-term conservation of *Hevea* germplasm

In situ and *ex situ* conservations are the two major strategies used in the conservation of plant genetic resources. *In situ* conservation, the conservation of diversity in its natural habitat, involves the designation, management and monitoring of the population at the location where it is currently found and within the community to which it belongs whereas *ex situ* conservation, collection of which the biodiversity is preserved outside its natural habitat, involves the sampling, transfer and storage of a population of a certain species away from the original location (Maxted et al., 1997). Because of a number of advantages, *the ex situ* method has been used to primarily conserve numerous plant genetic resources. In this method, plant diversity is safely preserved and concentrated in a small number of controlled places under consistent environmental conditions and is readily accessible to breeders.

An alternative solution to rubber tree *in situ* genetic conservation is the management of existing *ex situ* collections (Le Guen et al., 2009). Because of the ease of vegetative propagation by grafting, many *ex situ* collections of *Hevea* were established in various rubber producing countries. In Vietnam, the *ex situ* conservation of *Hevea* germplasm was established in 1985. This germplasm included the collection of local *Hevea* accessions, the introduction of *Hevea* clones from other countries. The majority of this germplasm collection was derived from the IRRDB'81 collection expedition in the Amazon forests of Brazil which is the primary center of diversity of the crop and the source of wild rubber trees. This germplasm was introduced into Vietnam in the form of budwood in 1984. All of the materials were sent to Lai Khe experimental station of the Rubber Research Institute of Vietnam (RRIV) in Binh Duong province, which is located in the traditional rubber growing belt. On receipt of the budwood, each accession was first multiplied by bud grafting and then planted in the field genebank for conservation in the form of source-bush garden (Fig. 3). This source-bush garden was laid out in randomized complete block design, in which each accession was represented in two replications of five trees with a planting distance of 1.5 m x 1.2 m. The genetic resources of *Hevea* germplasm conserved in Vietnam were showed in Table 1. The source-bush garden is cut back every year to maintain the conservation and also to generate budwood for various evaluation trials. Several preliminary field evaluation trials for most of *Hevea* germplasm accessions have been being established in several representative locations to evaluate their agronomical and morphological characteristics.

Fig. 3. The source-bush garden for *ex situ* conservation of *Hevea* germplasm in Vietnam

Genetic sources	Number of accessions
South America	**3082**
Acre (AC)	959
Mato Grosso (MT)	901
Rodonia (RO)	1116
Others	106
Africa	**38**
Ivory Coast	38
Asia	**422**
Vietnam	338
Malaysia	55
Indonesia	7
Sri Lanka	16
China	5
Cambodia	1
Total	**3542**

Table 1. The genetic resources of *Hevea* germplasm conserved in Vietnam

4. Characterization and evaluation of *Hevea* germplasm

In order to use the germplasm in breeding programs, it must be characterised and evaluated. There is often a delay between collection of germplasm and its evaluation, particularly for rubber trees because of the time required for them to reach maturity. Evaluation is useful if it considers the traits wanted by plant breeders. We are fortunate that our program of germplasm acquisition and evaluation is very closely linked to our program of *Hevea* breeding with the same people usually involving in both. Preliminary evaluation can help indicate those accessions that need more detailed evaluation, but those that appear not to be of immediate values should not be simply discarded.

4.1 Agronomical characteristics

Standard characterization and evaluation of germplasm collection may be routinely performed using different methods including traditional practices such as the use of descriptive lists of morphological characters. They may also involve evaluation of the agronomical performances under various environmental conditions. Understanding the nature and the magnitude of variability of important traits existing among plant genetic materials is vital for the effective utilization of such materials for breeding purposes. In rubber tree, high latex yield is always the exclusive objective of breeding programs. To this ultimate objective, many different factors are associated. For instance, the main components of productivity are the growth of the trunk determined during the immature period before the beginning of tapping, the resistance to various diseases and the tolerance to stress factors such as high altitude, low temperature, wind damage and moisture deficit.

In Vietnam, a part of *Hevea* germplasm, especially IRRDB'81 collection, has been agronomically evaluated in various clonal trials including arboreta and small scale clonal trials at different locations. In the view of latex production, IRRDB'81 collection exhibited very poor performance with an average latex yield of around 16% of the level of the currently developed Wickham clones after 5 years of tapping. This result was similar to that of other studies conducted on IRRDB'81 collections in Malaysia, Indonesia, Ivory Coast and China (Ramli et al., 2004; Aidi et al., 2002; Clément-Demange et al., 2002; Hu et al., 2002). Outstandingly, in the first three years of tapping, some IRRDB'81 accessions such as AC56/276, AC62/54, MT8/27, MT/I/2 and RO62/26 produced 30.0 – 45.0 gram/tree/tapping, or 102.7 – 153.8% of the production of the control clone (GT1). These accessions have been being used as parents in *Hevea* breeding programs. Considering geographical origins, accessions derived from Mato Grosso seemed to be better yielder than those from Acre and Rondonia. This might indicate their better adaptability to the experimental areas where the climate featured by a distinct dry spell of six months is similar to that of the original region which is known to have a dry spell of three to four months annually (Chevallier, 1988). A large number of IRRDB'81 accessions showed very good growth performance with girth at opening ranging from 59.9 – 74.0 cm, or 124.8 – 133.5% of that of the control clone. Remarkably, some IRRDB'81 accessions such as MT29/68, RO24/58 and RO32/104 had significantly higher girth than the control when planted in the highland area which was considered as a non-traditional rubber region in Vietnam. The superiority of these accessions will be of considerable value in advanced breeding programs.

Recently, IRRDB'81 collection has been considered as an important source for timber selection and rubber wood production. The average wood volume of the IRRDB'81 collection conserved in Vietnam was initially recorded, yielding 1.48 m³/tree at the age of 21 years, 43.4% higher than the Wickham population. Among these, several accessions appeared to be the best with clear bole volume ranging from 1.55 to 2.07 m³/tree, which could be considered as suitable clones for timber production purposes. Similarly, several high timber yield accessions of the IRRDB'81 collection in Indonesia and Malaysia were also reported. At the age of 13 years, a total of 28 accessions in Indonesia and 20 accessions in Malaysia were selected for timber yield with clear bole volume at a range of 0.90 to 2.56 m³/tree (Aidi et al., 2002) and 1.0 to 1.6 m³/tree (Ramli et al., 2004), respectively.

It is known that unlike other clonally multiplied species, *Hevea* is not affected by viral diseases (Simmonds, 1989). Other diseases which are considered as economic importance are *Gloeosporium* leaf disease (*Colletotrichum gloeosporioides* Pen. Sacc.), pink disease (*Corticium salmonicolor* Berk. & Br.), powdery mildew (*Oidium heveae* Stein.), *Corynespora* leaf fall (*Corynespora cassiicola* Berk. & Curt. Wei.), *Phytophthora* leaf fall (*Phytophthora* sp.) and SALB (South American Leaf Blight - *Microcyclus ulei* P. Henn von Arx.). Among these diseases, *Corynespora* leaf fall and SALB are the most important in rubber plantations. Wycherly (1969) noted that the clonal and location specificity was evident towards resistance to these diseases. Differences in the level of resistance to important diseases have been observed in IRRDB'81 collections from different countries. In India, a large number of IRRDB'81 accessions were showed to be resistant to powdery mildew (140 accessions) and *Corynespora* leaf fall (70 accessions) (Varghese et al., 2002). Similarly, a total of 21 accessions in IRRDB'81 collection are resistant to powdery mildew in bush-wood garden in China (Huang et al., 2002). The resistance to SALB was observed on 298 accessions from *ex situ* germplasm collection planted in both French Guyana and Brazil, of which the accessions from Acre and Rondonia were the most resistant (Le Guen et al., 2002). In Vietnam, it seemed that Wickham and IRRDB'81 accessions showed no significant difference in susceptibility to pink disease, powdery mildew and *Gloeosporium* leaf diseases (Lam et al., 2002). In general, evaluations of IRRDB'81 collection for biotic and abiotic stresses are in progress in major rubber growing countries.

4.2 Genetic diversity based on biochemical and molecular markers

In conventional plant breeding, many morphological traits have been used as markers for genetic analyses and cultivar identification, but specific genetic information on Mendelian traits are rare in *Hevea*. In contrast to traditional practices, genetic characterization refers to the description of the attributes that follow Mendelian inheritance or involve specific DNA sequences. In this way, biochemical assays, which detect differences between isozymes, or molecular markers were applied to determine the genetic diversity of the germplasm. In addition, the development of molecular and biochemical markers help researchers not only to identify genotypes, but also to assess and exploit the genetic variability (Whitkus et al., 1994). Insights into the relative genetic diversity among *Hevea* collections would be useful in *Hevea* breeding as well as *ex situ* conservation of *Hevea* genetic resources. The commercial value associated with identifying useful traits, especially yield and growth, would create a direct value in genebanks, ensuring long-term preservation of a collection. Moreover, *Hevea* germplasm characterization using molecular and biochemical markers will contribute to the knowledge of genetic relationships not only among wild accessions but also between accessions of wild and cultivated gene pool, and hence help to facilitate the breeding programs. In Vietnam, significant progress has been made in evaluation and characterization of *Hevea* germplasm by applying genetic markers. Among several efficient methods to reveal the genetic variability within and among plant populations, the most widely applied methods are isozyme electrophoresis and random amplified DNA polymorphism (RAPD). Both markers are useful to analyze genetic diversity of *Hevea* germpasm, and to select good *Hevea* clones for future breeding or cultivation purposes.

4.2.1 Isozymes marker

As first described by Hunter and Markert (1957), isozymes were defined as the different molecular forms in which proteins may exist with the same enzymatic specificity (Buth, 1984). This means that different variants on the same enzymes have identical or similar functions and are present in the same individual. Isozyme had played a minor role in research on plant biochemistry until genetic polymorphism for isozymes within the same population was discovered in 1966 (Stebbins, 1989; Wendel, 1989). In the 1980s, analysis of isozymes was developed at CIRAD (French Agricultural Research Centre for International Development) with 13 polymorphic isozymic systems to formulate a diagnostic kit associated with a clonal identification database. This kit has proved to be able to differentiate a large set of cultivated clones (Leconte et al., 1994). Since then, isozymes have been used as genetic markers for identification of rubber tree cultivars, genetic diversity analysis, controlling progenies issued from hand pollination and reproductive biology (Chevallier, 1988; Leconte et al., 1994; Paiva et al., 1994; Sunderasan et al., 1994). The first study on using isozyme markers for a precise understanding of the genetic diversity of the wild *Hevea* germplasm was carried out using ten isozyme markers on a set of 263 accessions from the IRRDB'81 collection (Chevallier, 1988).

Isozyme markers were firstly used in Vietnam to identify and confirm rubber clones in budwood gardens. Moreover, it was also used in *Hevea* breeding for hybrid genealogical legitimacy and genetic diversity research of *Hevea* germplasm. A total of 12 isozyme systems were used in studying genetic diversity of the IRRDB'81 *Hevea* germplasm in Vietnam. Banding patterns of representative IRRDB'81 accessions generated by isozyme electrophoresis are shown in Figure 4.

The study was performed on both IRRDB'81 collection and Wickham population with 117 accessions from 15 districts of the states of Acre, Rondonia and Mato Grosso of Brazil and 24 Wickham's clones (Fig. 5 and Table 2). The quantity of accessions sampled for each district was more or less proportional to the quantity of accessions currently conserved for the district. The result of isozymatic analysis showed that a total of 60 alleles were detected (Table 2). Out of them, 60 alleles were detected in IRRDB'81 accessions and 26 alleles in Wickham population. The result showed that the alleles detected in Wickham population were also detected in the IRRDB'81 collection, many new alleles were found in the IRRDB'81 collection only, thus underlining the genetic enrichment provided by the wild *Hevea* collections. Among IRRDB'81 collection, Acre accessions had the largest number of polymorphic alleles (51/60 alleles), followed by the Rodonia (45/60 alleles) and Mato Grosso accessions (45/60 alleles). This result revealed that Mato Grosso was obviously less polymorphic than Rondonia because Mato Grosso and Rondonia groups had the same number of detected polymorphic alleles but the volume of samples of the Mato Grosso groups was bigger than that of the Rondonia groups. The result also showed the significant polymorphic differences among the districts. The number of alleles detected in the districts Brasileia, Taurauca of Acre and district Cartriquacu of Mato Grosso was high in spite of the small volume of the samples whereas the number of alleles detected in districts Itanba and Aracatuba of Mato Grosso was quite low.

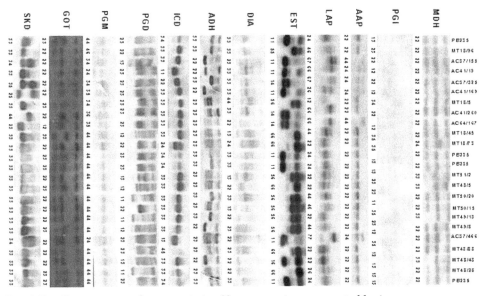

Fig. 4. Banding patterns of representative *Hevea* accessions generated by isozyme electrophoresis. Twelve isozyme systems used in studying genetic diversity of the IRRDB'81 *Hevea* germplasm had very different banding patterns. MDH, malate dehydrogenase; PGI, phospho glucose isomerase; AAP, alanyl amino peptidase; LAP, leucine amino peptidase; EST, esterase; DIA, diaphorase; ADH, alcohol dehydrogenase; ICD, isocitrate dehydrogenase; PGD, phosphogluconase dehydrogenase; PGM, phosphoglucomutase; GOT, glutamate oxaloacetate transaminase; SKD, shikimate acid

The study affirmed that the genetic base of *Hevea* germplasm in Vietnam, especially IRRDB'81 collection, was prosperous and diversified. Because of the high genetic variability level, this collection would contribute effectively to the long term progress of *Hevea* breeding and selection program in the country. In contrast, the Wickham population showed a low level of genetic variability, which is the consequence of the oriented selection through many years in a narrow geographical origin.

The result of cluster analysis based on isozymes database revealed the noticeable relationship between genetic clusters (Fig. 6). According to the genetic distance between the accessions, Acre accessions and Rondonia accessions were close to each other, meanwhile the genetic distance between Acre or Rondonia accessions were far from Mato Grosso accessions, except that the accessions from Vila Bella district of Mato Grosso (MT/VB) were close to those of Rondonia. Among IRRDB'81 accessions, Mato Grosso population was relatively close to Wickham collection based on isozyme analysis. In general, the genetic distance between the accessions conformed to the geographical origins of *Hevea*. However, several accessions of Arce and Rondonia were not separated distinctly using isozyme electrophoresis although they were distributed widely according to the geographical origins.

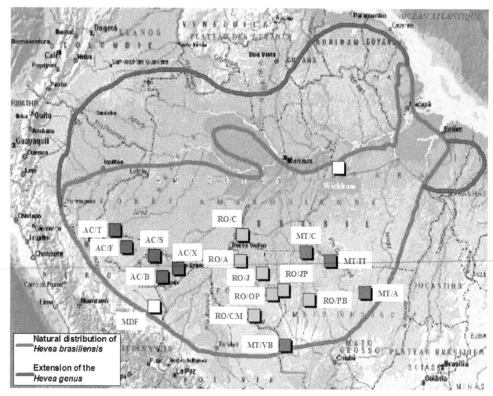

Fig. 5. Geographical origins of *Hevea* IRRDB'81 collection

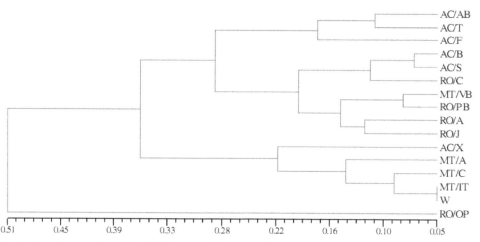

Fig. 6. Dendrogram based on Nei's genetic distance between IRRDB'81 collection and Wickham population.

Genetic resources	Number of accessions	MDH	PGI	AAP	LAP	EST	DIA	ADH	ICD	PGD	PGM	GOT	SKD	Total of alleles
IRRDB'81	117	4	5	5	7	7	5	2	5	3	6	4	7	60
Acre	58	4	5	3	6	6	5	2	4	3	6	3	4	51
Assis-brasil	2	1	2	2	3	2	1	2	2	2	2	1	3	23
Brasileia	9	3	3	2	3	4	3	2	4	3	5	2	3	37
Feijo	15	3	3	2	4	3	4	2	4	3	4	2	4	38
Sena Madureira	21	2	5	2	4	4	3	2	4	3	6	3	4	42
Tarauaca	8	3	3	2	3	2	2	2	3	2	4	1	3	30
Xapuri	3	1	2	1	3	2	2	2	1	3	4	3	2	26
Mato Grosso	32	3	5	1	6	5	4	2	4	3	4	3	5	45
Aracatuba	9	2	4	1	3	3	3	2	3	3	1	2	1	28
Cartriquacu	7	2	5	1	5	5	4	2	3	3	3	3	1	37
Itanba	13	2	3	1	4	3	2	2	3	2	2	3	2	29
Vila Bela	3	1	3	1	1	3	1	1	2	2	3	1	4	23
Rondonia	27	3	5	4	4	4	2	2	4	2	5	3	7	45
Ariquemes	5	1	2	3	2	3	2	2	2	2	5	2	4	30
Calama	18	2	5	1	3	4	2	2	4	2	5	3	6	39
Jaru	2	1	1	1	1	2	1	2	2	2	2	1	3	19
Ouro Preto	1	1	1	2	1	2	1	1	2	1	2	1	2	17
Pimenta Bueno	1	2	2	1	1	2	1	1	1	2	2	1	2	18
WICKHAM	24	2	3	1	3	3	1	2	4	3	1	2	1	26
TOTAL	141	4	5	5	7	7	5	2	5	3	6	4	7	60

* MDH, malate dehydrogenase; PGI, phospho glucose isomerase; AAP, alanyl amino peptidase; LAP, leucine amino peptidase; EST, esterase; DIA, diaphorase; ADH, alcohol dehydrogenase; ICD, isocitrate dehydrogenase; PGD, phosphogluconase dehydrogenase; PGM, phosphoglucomutase; GOT, glutamate oxaloacetate transaminase; SKD, shikimate acid

Table 2. Genetic variability of *Hevea* germpalsm based on isozymes

The results of isozymes analysis indicated that the *Hevea* germplasm conserved in Vietnam is very diversed. This characterisation would help to utilize the new genetic resources more effectively in *Hevea* breeding programs. The combination of morphological characterization and isozyme markers could help breeders to constitute a core collection of *Hevea* IRRDB'81 germplasm to ensure the conservation of the genetic variability. In addition, maintaining the genetic variability in *Hevea* germplasm would help to reduce gene erosion. Moreover, isozyme markers could be used as an assistant tool to orient a long term plan to advance heterosis to improve Wickham materials based on the recombination between Wickham clones and IRRDB'81 accessions. However, isozyme-based analysis is limited by the rather small number of marker loci available and a general lack of polymorphism for these loci. In addition, the analysis has to be carried out near the field sites owing to the fragility of the isozymes to varied temperatures or otherwise the samples need to be freeze-dried and transported to the laboratory. In spite of such limitations, isozyme was still a helpful marker to evaluate the genetic variability of the *Hevea* germplasm in Vietnam.

4.2.2 RAPD marker

The random amplified polymorphic DNA (RAPD) technique, first described by Williams et al. (1990), despite some limitations, has provided a useful approach for evaluating population's genetic differentiation, particularly in species that are poorly genetically known (Silva and Russo, 2000; Nybom, 2004). Recently, a large number of studies have pointed out that DNA-based markers, such as RAPD, were superior to isozymes in detecting genetic diversity (Garkava et al., 2000; Matos et al., 2001; Ochiai et al., 2001; Sharma et al., 2008). It is known that isozymes represent allelic expression of the same locus, while DNA fragments produced by RAPD are independent genetic markers (Ochiai et al., 2001) with a lower proportion of non-neutral markers than formerly (Bartish et al., 2000). Hence, isozyme and RAPD analyses often give discordant patterns, suggesting the importance of using multiple molecular marker systems in studies of population structure (Wendel and Doyle, 1998; Bartish et al., 2000; Lebot et al., 2003). RAPD marker was also used to evaluate the levels of gene flow between species (Arnold et al., 1991) and detection of gene introgression in various plant species (Waugh et al., 1992; Orozco-Castillo et al., 1994; Gomez et al., 1996). In rubber tree, RAPD has become a useful maker for investigating genetic diversity within and between *Hevea* populations, especially the IRRDB'81 collection (Varghese et al., 1997; Venkatachalam et al., 2002; Lam et al., 2009). Moreover, this marker was also used to identify a dwarf genome-specific marker (Venkatachalam et al., 2004) or certain homology to proline-specific permease gene (Venkatachalam et al., 2006) in rubber tree. The accumulated data on *Hevea* RAPD analysis from different accessions give information on genetic relations and *Hevea* origin, and provide the initial basis for clonal distinction and germplasm evaluation of agronomical interest. Therefore, the data can also be used in *Hevea* improvement programs.

In Vietnam, RAPD was firstly used to study genetic diversity of *Hevea* germplasm (Lam et al., 2009). The study was performed on IRRDB'81 collection with 59 accessions from 13 districts of the states of Acre, Rondonia and Mato Grosso of Brazil (Fig. 5). Using 6 oligonucleotide primers, the percentage of polymorphic loci calculated for individual districts ranged from 15.38% in Assis-Brasil district to 70.77% in Sena Madureira district of Acre, which totally had 10 and 46 polymorphic banding patterns, respectively (Table 3, Fig. 7). Although the sample sizes might have certain effects on the extent of the polymorphism of various districts, in the cases of Assis-Brasil of Acre, Ariquemes of Rondonia, and Vila Bela of Mato Grosso, they were quite different in the extent of polymorphism with the same sample sizes. In addition, the Jiparana district of Rondonia was very polymorphic regardless of its small sample size.

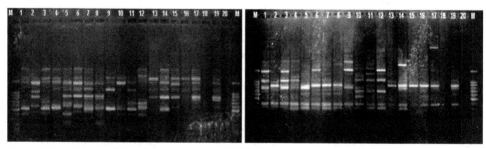

Fig. 7. DNA fingerprints of representative IRRDB'81 accessions generated by primer A18 (left) and OPB-12 (right)

State/District	No. of accessions	No. of total/polymorphic band patterns	Observed number of alleles	Mean heterozygosity	Shannon index	Genetic distance*
Acre	25					
Assis-brasil	2	41/10	1.154	0.064	0.093	0.139
Brasileia	4	52/37	1.569	0.223	0.328	0.299
Feijo	6	56/42	1.646	0.240	0.355	0.282
Sena Madureira	9	58/46	1.708	0.246	0.368	0.256
Tarauaca	4	56/32	1.492	0.196	0.287	0.225
Mato Grosso	21					
Aracatuba	6	51/38	1.585	0.227	0.333	0.264
Cartriquacu	5	51/38	1.585	0.222	0.328	0.306
Itanba	8	56/42	1.646	0.231	0.344	0.264
Vila Bela	2	46/21	1.323	0.134	0.195	0.296
Rondonia	13					
Ariquemes	2	45/16	1.246	0.102	0.149	0.216
Calama	4	53/36	1.554	0.217	0.319	0.290
Jaru	4	51/39	1.600	0.239	0.348	0.323
Jiparana	3	54/44	1.677	0.264	0.389	0.469
Total/Mean	59	65/62	1.522	0.200	0.295	0.279

* Genetic distance is mean genetic distance among accessions in individual district

Table 3. Patterns of genetic diversity of *Hevea* IRRDB'81 collection based on RAPD analysis

In general, the IRRBD'81 collection conserved in Vietnam showed the high level of genetic diversity detected by RAPD. In fact, the mean values of heterozygosity or Nei's genetic diversity (Nei, 1978) within individual districts varied from 0.064 to 0.264 over 65 loci with the average of 0.2 across the districts (Table 3). This parameter differed substantially among 13 districts studied. The Jiparana district of Rondonia showed the highest estimated heterozygosity over 62 polymorphic loci, whereas the Assis-Brasil district of Acre showed the lowest one. In spite of high standard errors of mean heterozygosities, probably due to the small sample sizes, recorded in Assis-Brasil (Acre), Ariquemes (Rondonia), and Vila Bela (Mato Grosso) populations, the remarkable variations of mean heterozygosity clearly showed differences in genetic variability among 13 districts. The average degree of diversity within individual districts using Shannon's diversity index (Shannon and Weaver, 1949) was 0.296 and ranged from 0.093 for the Assis-Brasil district of Acre to 0.389 for the Jiparana district of Rondonia (Table 3). Shannon index was correlated strongly with the percentage of polymorphic loci in a district. In fact, the districts with high Shannon's diversity index also exhibited the high percentages of polymorphic loci (Table 3). Similar to mean heterozygosity, differences in values of Shannon's diversity index also showed genetic differentiation among the districts. The largest average genetic distance among accessions within the districts was detected in the Jiparana district of Rondonia and the smallest was found in the Assis-Brasil district of Acre, of which the average genetic distance values were 0.469 and 0.139, respectively (Table 3). According to previous studies, high genetic diversity is usual in IRRDB'81 accessions (Chevallier, 1988; Besse et al., 1994; Lekawipat et al., 2003). All the accessions had unique RAPD genotypes. Nei's genetic distance values between pairs of districts ranged from 0.046 for Catriquacu and Itanba of Mato Grosso to 0.304 for Tarauaca of Acre and Aracatuba of Mato Grosso (Table 4). The dendrogram constructed by UPGMA cluster analysis showed that *Hevea* IRRDB'81 collection of 13 different districts were in five clusters with Ariquemes of Rondonia quite different from the others (Fig. 8).

Among the districts of Mato Grosso, Vila Bela was in the same cluster with the districts from Rondonia (Jaru and Jiparana), meanwhile other districts (Itanba, Catriquacu and Aracatuba) were grouped into one cluster. This showed that Vila Bela was quite different from other districts of Mato Grosso, which was also noted by other researchers using RFLP markers (Besse et al., 1994). The distribution of those districts in genetic cluster analysis seemed to conform to geographical origins of *Hevea* IRRDB'81 collection, except Calama district of Rondonia.

District	AC/AB	AC/B	AC/F	AC/S	AC/T	MT/A	MT/C	MT/IT	MT/VB	RO/A	RO/C	RO/J	RO/JP
AC/AB		0.147	0.201	0.146	0.160	0.297	0.244	0.236	0.202	0.286	0.188	0.236	0.266
AC/B	0.147		0.106	0.107	0.155	0.190	0.135	0.161	0.216	0.275	0.135	0.160	0.164
AC/F	0.201	0.106		0.105	0.145	0.125	0.115	0.129	0.213	0.230	0.107	0.144	0.150
AC/S	0.146	0.107	0.105		0.190	0.153	0.151	0.146	0.158	0.239	0.103	0.164	0.169
AC/T	0.160	0.155	0.145	0.190		0.304	0.227	0.234	0.193	0.236	0.230	0.205	0.212
MT/A	0.297	0.190	0.125	0.153	0.304		0.094	0.086	0.296	0.295	0.171	0.191	0.183
MT/C	0.244	0.135	0.115	0.151	0.227	0.094		0.046	0.202	0.286	0.125	0.145	0.141
MT/IT	0.236	0.161	0.129	0.146	0.234	0.086	0.046		0.189	0.289	0.147	0.145	0.147
MT/VB	0.202	0.216	0.213	0.158	0.193	0.296	0.202	0.189		0.233	0.123	0.146	0.149
RO/A	0.286	0.275	0.230	0.239	0.236	0.295	0.286	0.289	0.233		0.222	0.189	0.205
RO/C	0.188	0.135	0.107	0.103	0.230	0.171	0.125	0.147	0.123	0.222		0.137	0.172
RO/J	0.236	0.160	0.144	0.164	0.205	0.191	0.145	0.145	0.146	0.189	0.137		0.097
RO/JP	0.266	0.164	0.150	0.169	0.212	0.183	0.141	0.147	0.149	0.205	0.172	0.097	

Table 4. Nei's genetic distance among 13 districts of *Hevea* IRRDB'81 collection

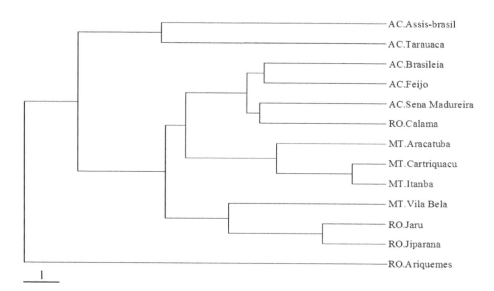

Fig. 8. Dendrogram of UPGMA cluster analysis based on Nei's genetic distance

Analysis of molecular variance (AMOVA) for 59 accessions of 13 districts revealed that the interpopulation (interdistrict) variation accounted for only 14.1% of the total genetic variance while intrapopulation (within district) variation accounted for 85.9% (Table 6). The results indicated that the majority of genetic diversity was within district variation (85.9%). The results suggests that there might be a certain gene flow among the districts, possibly owing to the species' outcrossing as a mode of reproduction and dispersion of seeds by a network of rivers in Amazon basin. However, it should be treated with caution due to small samples brought into the study.

Source of variation	df	SS	CV	%Total
Among districts	12	198.6	1.57	14.1
Within districts	46	439.6	9.56	85.9

Significant at P=0.001

Table 5. Summary of the AMOVA analysis

The IRRDB'81 *Hevea* germplasm in Vietnam exhibited large genetic variability by RAPD markers. The intradistrict source accounted for most of the genetic variation of the germplasm. Based on the genetic distance analysis, the collection could be classified into five groups which could help in planning crosses based on genetic distance in the hope of looking for heterosis and maintaining the abundant genetic diversity. The RAPD markers could also help in checking the genetic variability of the *Hevea* breeding program. Moreover, in association with the agronomical characteristics, morphological traits and isozymes analysis, RAPD markers are now suitable tools for genetic diversity studies of *Hevea* germplasm and can be useful for accumulation and management of genetic-breeding resources of *Hevea brasiliensis*.

5. Utilization of *Hevea* germplasm in Vietnam

One of the major objectives of conservation of *Hevea* genetic resources is to make genetic diversity available for immediate or future use. The widest possible range of the genetic diversity has to be conserved in order to meet future, as yet unknown, needs. *Hevea* germplasm conservation program in Vietnam is expected to promote and facilitate the use of conserved materials through the maintenance of healthy and readily accessible and adequately characterised/evaluated materials, and proper documentation of the relevant information. Evaluation data of the agronomic performances and the morphological characteristics gathered during cultivation of the accessionshave been being recorded continuously using a specifically constructed program (Fig. 9). Currently, a total of about 3,500 *Hevea* accessions have been collected and *ex situ* conserved. This germplasm comprises of three main genetic resources: the Amazonian (A) (most of which belong to IRRDB'81 collection collected in the Amazonian habitats of the genus), the Wickham (W), and the Wickham x Amazonian (WA) resources. The majority of this germplasm were derived from the IRRDB'81 collection with a total of 2,999 accessions, each of which is a clone derived from originally collected seedlings. Most of them have been evaluated for the agronomical and morphological characteristics. In the view of the limitations of the agronomical and morphological traits, isozyme and RAPD markers were used to analyze the genetic diversity and structure of the IRRDB'81 collection for more effective utilization of the germplasm in *Hevea* breeding programs in Vietnam.

Since the IRRDB'81 collection exhibited very poor profiles in agronomical characteristics, especially latex productivity, the chance for direct use of this collection for latex purposes seemed to be impractical although certain accessions could be planted for timber purposes. Regarding widening the genetic base for genetic improvement, several promising IRRDB'81 accessions have been included in hand pollination program in RRIV since 1997. Based on the agronomical and morphological traits as well as the genetic diversity analysis, recently, many attempts have been made to enlarge the genetic base of *Hevea* breeding materials by polycrossing among different genetic resources. In this way, many crosses between maternal

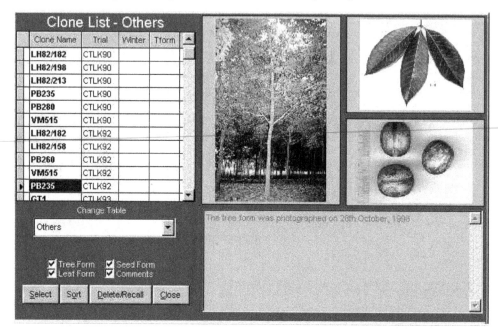

Fig. 9. Genebank documentation program in conservation of *Hevea* germplasm

W and paternal A accessions have been preferably made and the progenies of which are in various phases of evaluation with the expectation that the W x A progenies could combine the good agronomical characteristics from parents while optimizing the genetic variability in this population. In fact, Amazonian accessions were crossed with Wickham high yielding and good set fruit clones such as PB260 and RRIC110, and progenies derived from these hand pollination crosses were disbudded into fields of early selection trials for evaluation of agronomical performances such as latex yield, growth and diseases incidence. In general, these progenies exhibited rather good in girth but very poor in latex productivity. Most of the progenies produced a very little or negligible amount of latex; this agreed closely with the previous finding in Ivory Coast for the progenies from W x A crosses (Clement-Demange et al, 1990). However, some progenies had the production of 1.9 – 3.2 gram/tree/tapping or 106 – 132% of the production of the control clone (i.e. PB 260), and a large number of the progenies showed very good in growth performance with girth at 34 months old after planting ranging from 15.6 – 21.2 cm, or 101 – 161% of that of the control clone. These progenies are being further tested in the small scale clonal trials and would be included as parents in future breeding programs. This gives a way for the opportunity of genetic improvement in breeding programs, particularly in latex productivity of the *Hevea* IRRDB'81 collection. In diseases incidence, all of the progenies exhibited varying degrees of infection to powdery mildew. The progenies derived from different paternal accessions showed significant differences in susceptibility to powdery mildew. For instance, the progenies derived from AC6/23 and AC35/114 were more susceptible to powdery mildew than other progenies. In contrast, the progenies derived from RO44/268 and RO44/71 seemed to be lightly susceptible to powdery mildew in comparison to other progenies. This result has contributed to the development of clonal disease resistance by genetic recombination using IRRDB'81 collection as paternal clones in *Hevea* breeding programs.

The results obtained so far can be considered as a basis to continue combining Wickham and IRRDB'81 genetic resources in breeding programs at RRIV. The W x A progeny population provides a valuable source for selecting multi-clones recommended for developing rubber cultivation in non-traditional regions in Vietnam.

6. Problems and challenges

6.1 Problems

Because the genetic structure of natural *H. brasiliensis* populations is based upon both hydrographical network patterns and long-range isolation by distance, it is likely that the known genetic diversity represents only part of its true natural diversity (Le Guen et al., 2009). In order to enlarge the genetic resources of *H. brasiliensis*, it is necessary to conduct additional expeditions in other areas that were not yet covered previously, such as Amazonian basin in Colombia, Peru and Bolivia, and the Brazilian states of Pará and Amazonas. Besides, since a very small part of the diversity of *Hevea* germplasm has been collected and conserved in Vietnam, a much greater diversity of the germplasm should be imported from other countries such as Brazil, Malaysia and Ivory Coast. Moreover, it is urgent need to duplicate the *Hevea* germplasm accessions, particularly the IRRDB'81 collection, in all IRRDB country members to prevent the loss of accessions and to increase the genetic resources as raw materials in *Hevea* breeding programs. Additionally, molecular tools may contribute in *ex situ* conservation of *Hevea* germplasm to the sampling, management and development of "core" collections as well as the utilization of genetic diversity. However, the use of such molecular tools is limited in the rubber growing countries including Vietnam due to their cost.

Rubber tree is traditionally propagated through bud grafting on unselected seedlings, which maintains intraclonal heterogeneity for vigour and productivity. Therefore, variation among a bud-grafted population is significant and can influence the growth and productivity levels. Therefore, a great improvement may be expected by using *in vitro* micropropagation. *In vitro* techniques have currently made a commercial impact in rubber, and their propagation systems can circumvent the influence of the stock-scion interactions in *Hevea* clones (Priyadarshan, 2007). Recently, there has been an increasing interest in the induction of somatic embryogenesis in rubber trees. However, successful somatic embryo formation and plant regeneration have been reported by a few researchers in different countries using limited genotypes of *Hevea*. In addition, the frequency of somatic embryo induction was found to be very low and non-synchronous, its germination remains very difficult and thus *Hevea* embryogenic system needs further investigation. There has been no large scale commercial application of tissue culture techniques for mass propagation of clonal *Hevea* as yet, either by microcutting or by somatic embryogenesis. However, there is sufficient progress at the research level to suggest that tissue culture of *Hevea* can and should be further developed.

6.2 Challenges

At the scientific and technical levels, challenges are posed by genetic erosion, genetic vulnerability and utilization. Genetic erosion is defined as "the loss of genetic diversity, including the loss of individual genes, and the loss of particular combinations of genes (i.e. of gene-complexes) such as those manifested in locally adapted landraces" (FAO, 1997). There is no consensus on the optimal balance of *in situ* and *ex situ* conservation methods to combat genetic erosion (Fraleigh, 2006). A gradual erosion of the genetic variability of *Hevea* in all of natural rubber plantations was realized in the 1970s. This erosion occurred because most of *Hevea* clones in cultivation were derived from the few surviving seeds collected by Henry Wickham. Therefore, widening the genetic base of *Hevea* in production was seen as a prerequisite to generate new clones with new and valuable characteristics with regard to productivity, disease resistance, and tolerance to many environmental conditions.

Genetic vulnerability was described as "the condition that results when a widely planted crop is uniformly susceptible to a pest, pathogen or environmental hazard as a result of its genetic constitution, thereby creating a potential for widespread crop losses" (FAO, 1996). It is known that genetic vulnerability pertains to the level of the crop genetic diversity actually being used. Because of the very narrow genetic base in the commercially cultivated *Hevea* clones, the commercial rubber cultivation, due to their genetic vulnerability, is under a constant threat of attack by native as well as exotic diseases and insect pests. The changes in the weather parameters due to the increasing trend in climate change have further complicated the above issues. Climate change, which is clearly felt in the traditional rubber growing regions of Vietnam, may possibly alter the host-pathogen interactions. This will lead to the emergence of otherwise minor disease, and *Corynespora* leaf fall disease may represents this scenario. This pathogen is rapidly progressing into new areas, thus highlighting the need for stronger and advanced resistance breeding approaches.

It is known that the potential uses and values of *Hevea* genetic resources need to be understood by characterizing, evaluating and documenting them. Methods still need to be developed to improve and facilitate productive utilization. Although biotechnological methods are now increasingly available to facilitate productive utilization of *Hevea* germplasm, not all countries have the capacity to use such new technologies.

Another set of challenges is posed for taking action. For instance, how the necessary cooperation can be organized among countries and among disciplines, particularly in order to link the conservation and the use of *Hevea* genetic resources, and how the resources which need to address these issues can be mobilized. Besides, cryogenic preservation of endangered seedling trees is yet another important aspect to be looked into urgently.

7. Conclusions and prospects

Vietnam has received a large share of the *Hevea* germplasm collection of the 1981 IRRDB expedition. A total of 2,999 wild accessions belonging to three states of Brazil *viz.,* Acre, Rondonia and Mato Grosso are being conserved. Systematic efforts are underway for conservation, characterization, utilization and documentation of these valuable genetic resources. In general, the genetic base of *Hevea* in Vietnam which is prosperous and diversified has been contributing effectively to the long term progress of *Hevea* breeding program in the country. As management of the germplasm is a herculean task, IRRDB'81 accessions have been evaluated in a phased manner. Studies in different sets of this germplasm have been carried out since 1985 onward to access the extent of variability present in the collection for various agronomical characteristics such as latex yield, girth, wood volume, diseases incidence, biotic and abiotic stresses, in order to exploit them in the improvement programs. A large number of IRRDB'81 accessions are now in various evaluation stages. The evaluated IRRDB'81 collection displayed unimproved characteristics of a wild population and was far inferior to the Wickham clones in agronomical performances, especially latex productivity. However, this germplasm had a much broader genetic variability which can help in broadening the narrow genetic base of cultivated rubber and also in developing location specific rubber clones for cultivation in the marginal and non-traditional rubber regions of the country.

The genetic parameters obtained from isozyme and RAPD analyses indicated that the *Hevea* germplasm conserved in Vietnam exhibited large genetic diversity. The biochemical and molecular markers have also shown to be the effective techniques for breeders to manipulate the *Hevea* germplasm. These markers could be used to select the parents with far genetic distance aiming at enlarging the genetic variability in their progenies and could help in checking the genetic variability of the *Hevea* breeding program. Furthermore, information on the structure of genetic diversity could help establish global *Hevea* collections for long-term conservation with minimum maintenance activity, and help define working collections for medium-term targeted utilization and breeding purposes.

The conventional solution to the conservation of *Hevea* genetic resources has been the establishment of *ex situ* genebanks. *Ex situ* conservation is the only effective method for the long-term conservation of *Hevea* germplasm. *Ex situ* conservation may also represent a last resort for many species and varieties including *Hevea* that would otherwise die out as their habitats are destroyed. Moreover, the management of existing *ex situ Hevea* collections is an alternative solution to rubber trees *in situ* genetic preservation (Le Guen et al., 2009). The main benefit of *ex situ* conservation is to provide breeders with ready access to a wide range of genetic materials with useful traits. *Hevea* germplasms have their own share of problems. Although only a small proportion of *Hevea* genetic resources are actually used by breeders, partly because of the inadequate characterization of the accessions, the costs of characterizing, evaluating and cataloguing genetic resource materials need to be carefully considered. Because of the severe limitations faced by large germplasm collection, establishing a core collection of this germplasm is necessary to facilitate a speedy and more efficient evaluation, and to get a collection which is conserved better and exploited more effectively.

8. References

Aidi, D.; Sekar, W. & Irwan, S. (2002). Report on The Evaluation and Utilization of The 1981 IRRDB *Hevea* Germplasm in Indonesia, *IRRDB Joint Workshop on Breeding, Agronomy and Socioeconomy*, Malaysia & Indonesia, 28 August - 7 September, 2002

Arnold, M.L.; Buckner, C.M. & Robinson, J.J. (1991). Pollen-Mediated Introgression and Hybrid Speciation in Louisiana Irises. *Proc. Natl. Acad. Sci. USA*, 88, 1398-1402

Baldwin, J.J.T. (1947). *Hevea*: A First Interpretation. A Cytogenetic Survey of a Controversial Genus, With a Discussion of Its Implications to Taxonomy and to Rubber Production. *Journal of Heredity*, 38, 54-64, ISSN 0022-1503

Bartish, I.V.; Garkava, L.P., Rumpunen, K. & Nybom, H. (2000). Phylognetic Relationship and Differentiation Among and Within Populations of *Chaenomeles* Lindl. (Rosaceae) Estimated with RAPDs and Isozymes. *Theor. Appl. Genet.*, 101, 554–563

Baulkwill, W.J. (1989). The History of Natural Rubber Production. In: *Rubber*, C.C. Webster & W.J. Baulkwill (Ed.), 1–56, Longman Scientific and Technical, ISBN 0-470-40405-3, Essex, UK

Besse, P.; Seguin, M., Lebrun, P., Chevallier, M. H., Nicholas, D., & Lanaud, C. (1994). Genetic Diversity among Wild and Cultivated Populations of *Hevea brasiliensis* Assessed by Nuclear RFLP Analysis. *Theor. Appl. Genet.*, 88, 199–207, ISSN 0040-5752

Bos, H. & McIndoe, K.G. (1965). Breeding of *Hevea* for Resistance Against *Dothidella ulei* P. Henn. *J. Rubber Res. Inst. Malaya*, 19, 98–107

Buth, D.G. (1984). The Application of Electrophoretic Data in Systematic Studies. *Ann. Rev. Ecol. Syst.*, 15, 501–522, ISSN 0066-4162

Chevallier, M.H. (1988). Genetic Variability of *Hevea brasiliensis* Germplasm Using Isozyme Markers. *J. Nat. Rubb. Res.*, 3, 42-53, ISSN 0127-7065

Clément-Demange, A.; Chapuset, T. & Seguin, M. (2002). Report on The IRRDB 1981 *Hevea* Germplasm by CIRAD (France) Year 2002. *IRRDB Joint Workshop on Breeding, Agronomy and Socioeconomy*, Malaysia & Indonesia, 28 August - 7 September, 2002

Clément-Demange, A.; Legnate, H., Seguin, M., Carron, M.P., Le Guen, V., Chapuset, T. & Nicolas, D. (2000). Rubber Tree. In: *Tropical Plant breeding*, A. Charrier, M. Jacquot, S. Hamon & D. Nicolas (Ed.), 455-480, Collection Repères, ISBN 978-1-57808-144-8, Montpellier, France.

Clément-Demange, A.; Priyadarshan, P.M., Hoa, T.T.T. & Venkatachalam, P. (2007). *Hevea* Rubber Breeding and Genetics. In: *Plant Breeding Review*, 29, J. Janick (Ed.), 177-281, John Wiley & Sons, Inc., ISBN 9780470052419, New York, USA

Dean, W. (1987). *Brazil and the Struggle for Rubber*, Cambridge Univ Press, ISBN 0-521-33477-2, New York, USA

FAO. (1996). First Report on The State of The World's Plant Genetic Resources for Food and Agriculture. FAO, Rome, Italia, Available from http://www.fao.org/WAICENT/FAOINFO/AGRICULT/AGP/AGPS/Pgrfa/pdf/swrshr_e.pdf

FAO. (1997). First Report on The State of The World's Plant Genetic Resources for Food and Agriculture. FAO, Rome, Italia, Available from http://www.fao.org/WAICENT/FAOINFO/AGRICULT/AGP/AGPS/pgrfa/pdf/swrfull.pdf

Fraleigh, B. (2006) Global Overview of Crop Genetic Resources. In: *The Role of Biotechnology in Exploring and Protecting Agricultural Genetic Resources*, J. Ruane & A. Sonnino (Ed.), 21-32, FAO, ISBN 92-5-105480-0, Roma, Italia

Garkava, L.P.; Rumpunen, K. & Bartish, I.V. (2000). Genetic Relationships in *Chaenomeles* (Rosaceae) Revealed by Isozyme Analysis. *Sci. Hortic*, 85, 21–35

George, P.J. (2000). Germplasm Resources. In: *Natural rubber: Agromanagement and Crop Processing*, P.J. George & C.K. Jacob (Ed.), 47-58, Rubber Research Institute of India, Kottayam, India

Gomez, R.; Angel, F., Bonierbale, M.W., Rodriguez, F., Tohme, J. & Roca, W.M. (1996). Inheritance of Random Amplified Polymorphic DNA Markers in Cassava (*Manihot esculenta* Crantz). *Genome*, 39, 1039-1043

Gonçalves, P. de S.; Cardoso M. & Ortolani A.A. (1990). Origin, variability and domestication of *Hevea*. *Pesquisa Agropecuária Brasileira*, 25, 135-156, ISSN 0100-204X

Hoa, T.T.T. (2010). Overview of Vietnam Rubber Industry and Prospects of Development by 2020, *Proceedings of the 9thAsian Workshop on Polymer Processing in Vietnam*, pp. xxiii-xxvii, Hanoi, Vietnam, December 7-10, 2010

Hu, Y.; Zeng, X., Chen, H., Fang, J. & Huang, H. (2005). A Study Report on The Main Characters Evaluation of New Amazon *Hevea* Germplasm, *International Natural Rubber Conference*, pp. 50-54, Cochin, India, November 6-8, 2005

Huang, H.; Yu, D. & Zhou, J. (2002). Studies of The 1981 IRRDB *Hevea* Germplasm in China, *IRRDB Joint Workshop on Breeding, Agronomy and Socioeconomy*, Malaysia & Indonesia, 28 August - 7 September, 2002

Hunter, R.L. & Markert, C.L. (1957). Histochemical Demonstration of Enzymes Separated by Zone Electrophoresis in Starch Gels. *Science*, 125, 1294-1295

IRSG. (2009). *Latest Rubber Statistical Bulletin*. International Rubber Study Group, London, October – December, 2009, Available from http://www.rubberstudy.com/default. aspx?aspxerrorpath=/newsarticle.aspx.

IRSG. (2010). *Latest Rubber Statistical Bulletin*. International Rubber Study Group, London, April – June, 2010, Available from http://www.rubberstudy.com/default. aspx?aspxerrorpath=/newsarticle.aspx

Lam, L.V.; Lam, H.B., Hoa, T.T.T. & Trang, L. T.T. (2002). Status Report of The IRRDB'81 *Hevea* Germplasm in Vietnam. *IRRDB Joint Workshop on Breeding, Agronomy and Socioeconomy*, Malaysia & Indonesia, 28 August - 7 September, 2002

Lam, L.V.; Thanh, T., Chi, V.T.Q., Tuy, L.M. (2009). Genetic Diversity of *Hevea* IRRDB'81 Collection Assessed by RAPD Markers. *Mol. Biotechnol.*, 42, 292–298

Le Guen, V. ; Garcia, D., Mattos, C.R.R. & Clément-Demange, A. (2002). Evaluation of Field Resistance to *Microcyclus ulei* of A Collection of Amazonian Rubber Tree *(Hevea brasiliensis)* Germplasm. *Crop Breeding and Applied Biotechnology*, 2, 141-148, ISSN 1518-7853

Le Guen, V.; Doaré, F., Weber, C. & Seguin, M. (2009). Genetic Structure of Amazonian Populations of *Hevea brasiliensis* is Shaped by Hydrographical Network and Isolation by Distance. *Tree Genetics & Genomes*, 5, 673-683, ISSN 1614-2942

Lebot, V.; Herail, C., Gunua, T., Pardales, J., Prana, M., Thongjiem, M. & Viet, N. (2003). Isozyme and RAPD Variation Among *Phytophthora colocasiae* Isolates from South East Asia and The Pacific. *Plant Pathol.*, 52, 303–313

Leconte, A.; Lebrun, P., Nicolas, D. & Seguin, M. (1994). Electrophoresis Application to *Hevea* Clone Identification. *Plant Research and Dévelopment*, 1, 28-36

Lekawipat, N., Teerawatanasuk, K., Rodier-Goud, M., Seguin, M., Vanavichit, A., Toojinda, T. & Tragoonrung, S. (2003). Genetic Diversity Analysis of Wild Germplasm and

Cultivated Clones of *Hevea brasiliensis* Muell. Arg. by Using Microsatellite Marker. *J. Rubber Res.*, 6, 36–47, ISSN: 1511-1768

Majumder, S.K. (1964). Chromosome Studies of Some Species of *Hevea*. *J. Rubb. Res. Inst. Malaysia*, 18, 269-273.

Matos, M.; Pinto-Carnide, O. & Benito, C. (2001). Phylognetic Relationships Among Portuguese Rye Based on Isozyme, RAPD and ISSR Markers. *Hereditas*, 134, 229–236

Maxted, N.; Ford-Lloyd, B.V. & Hawkes, J.G. (1997). Complementary Conservation Strategies. In: *Plant Genetic Conservation: The In Situ Approach*, N. Maxted, B.V. Ford-Lloyd & J.G. Hawkes (Ed.), 20-55, Chapman & Hall, ISBN 0-412-63730-8, London, UK

Nei, M. (1978). Estimation of Average Heterozygosity and Genetic Distance from a Small Number of Individuals. *Genetics*, 89, 583–590.

Nicolas, D. (1985). Acquisition of *Hevea* Material Derived from Colombian Schultes Collections, *International Rubber Conference*, Kuala Lumpur, Malaysia, October 18-19, 1985

Nybom, H. (2004). Comparison of Different Nuclear DNA Markers for Estimating Intraspecific Genetic Diversity in Plants. *Molecular Ecology*, 13, 1143–1155

Ochiai, T.; Nguyen, V.X., Tahara, M. & Yoshino, H. (2001). Geographical Differentiation of Asian Taro, *Colocasia esculanta* (L.) Schott, Detected by RAPD and Isozyme Analyses. *Euphytica*, 122, 219–234

Ong, S.H. (1975). Chromosome Morphology at the Pachytene Stage in *Hevea brasiliensis*: A Preliminary Report, *Proceedings of the International Rubber Conference*, pp. 3-12, Kuala Lumpur, Malaysia

Onokpise, O.U. (2004). Natural Rubber, *Hevea brasiliensis* (Willd. ex A. Juss.) Müll. Arg., Germplasm Collection in The Amazon Basin, Brazil: A Retrospective. *Econ. Bot.*, 58, 544-555, ISSN 0013-0001

Orozco-Castillo, C.; Chalmers, K.J., Waugh, R. & Powell, W. (1994). Detection of Genetic Diversity and Selective Gene Introgression in Coffee Using RAPD Markers. *Theor. Appl. Gent.*, 87, 934-940

Orwa, C.; Mutua, A., Kindt, R., Jamnadass, R. & Simons, A. (2009). *Agroforestree Database: A Tree Reference and Selection Guide version 4.0*, Available from http://www.worldagroforestry.org/af/treedb/

Paiva, J.R.; Kageyama, P.Y., Vencorsky, R. & Contel, P.B. (1994). Genetics of Rubber Tree *Hevea brasiliensis* (Willd. ex Adr. de Juss.) Muell. Arg. 1. Genetic Variation in Natural Populations. *Silvae Genetica*, 43, 307-312

Priyadarshan, P.M. & Gonçalves P. de S. (2002). Use of *Hevea* gene pool in rubber tree (*Hevea brasiliensis* Muell. Arg) breeding. *The Planter*, 78, 123-138

Priyadarshan, P.M. (2007). Rubber. In: *Genome Mapping and Molecular Breeding in Plant: Tropical Crops*, C. Kole (Ed.)., 143-174, Springer-Verlag, ISBN 978-3-540-34537-4, New York, USA

Raemer, H. (1935). Cytology of *Hevea*. *Genetics*, 17, 193, ISSN 0016-6731

Ramli, O.; Masahuling, B., Nasaruddin M.M.A. & Zarawi, A.G. (2004). IRRDB 1981 Expedition: Harnessing Genetic Potential of *Hevea* Germplasm, *Proceedings of IRRDB Conference*, pp. 11-25, Kunming, China, September 7-8, 2004

Rands, R.D. & Polhamus, L.G. (1955). *Progress Report on The Cooperative Hevea Rubber Development in Latin America*, U.S. Dept. of Agriculture, Washington, USA

Schultes, R.E. (1977). A New Infrageneric Classification of *Hevea*. *Botanical Museum Leaflets of Harvard University*, 25, 243–257

Schultes, R.E. (1987). Studies in the Genus *Hevea* VIII. Notes on intrageneric variants of *Hevea brasiliensis* (Euphorbiaceae). *Econ. Bot.*, 41, 125-147

Schultes, R.E. (1990). A Brief Taxonomic View of the Genus *Hevea*. *MRRDB Monograph* 14, ISSN 0127-7782, Kuala Lumpur, Malaysia.

Seguin, M.; Flori, A., Legnate, H., Clément-Demange, A. (2003). Rubber Tree. In: *Genetic diversity of tropical crops*, P. Hamon, M. Seguin, X. Perrier & J.C Glaszmann JC (Ed.), 277–305, Science Publisher, Enfield, USA

Shannon, C. E. & Weaver, W. (1949). *The Mathematical Theory of Communication*, University of Illinois Press, ISBN 0-252-72546-8, Illinois, USA

Sharma, K.; Mishra, A.K. & Misra, R.S. (2008). A Simple and Efficient Method for Extraction of Genomic DNA from Tropical Tuber Crops. *Afr. J. Biotechnol.* 7, 1018–1022

Silva, E.P. & Russo, C.A.M. (2000). Techniques and Statistical Data Analysis in Molecular Population Genetics. *Hydrobiologia*, 420, 119–135

Simmonds, N.W. (1989). Rubber Breeding. In: *Rubber*, C.C. Webster & W.J. Baulkwill (Ed.), 85-124, Longman Scientific and Technical, ISBN 0-470-40405-3, Essex, UK

Stebbins, G.L. (1989). Introduction. In: *Isozymes in Plant Biology*, D.E. Soltis & P.S. Soltis (Ed.), 1-4, Dioscorides Press, Oregon, USA

Sunderasan, E.; Wickneswari, R., Aziz, M.Z.A. & Yeang, H.Y. (1994). Incidence of Self- and Cross-Pollination in Two *Hevea brasiliensis* Clones. J. Nat. Rubber Res., 9, 253-257, ISSN 0127-7065

Tan, H. (1987). Strategies in Rubber Tree Breeding. In: *Improving Vegetatively Propagated Crops*, A.J. Abbott & R.K. Atkin. (Ed.), 28-54, Academic Press, ISBN 0-120-41410-4, London, UK.

Varghese, Y.A.; Abraham, S.T., Mercy, M.A., Madhavan, J., Reghu, C.P., Rao, G.P., Ammal, S.L., Idicula, S.P. & Joseph, A. (2002). Management of The 1981 Germplasm Collection in India, *IRRDB Joint Workshop on Breeding, Agronomy and Socioeconomy*, Malaysia & Indonesia, 28 August - 7 September, 2002

Varghese, Y.A.; Knaak, C., Sethuraj, M.R. & Ecke, W. (1997). Evaluation of Random Amplified Polymorphic DNA (RAPD) Markers in *Hevea brasilinesis*. *Plant Breeding*, 116, 47-52

Venkatachalam, P.; Priya, P., Gireesh, T., Saraswathyamma, C.K. & Thulaseedharan, A. (2006b). Molecular Cloning and Sequencing of a Polymorphic Band from Rubber Tree [*Hevea brasiliensis* (Muell.) Arg.]: The Nucleotide Sequence Revealed Partial Homology with Proline-Specific Permease Gene Sequence. *Current Science*, 90, 1510-1515

Venkatachalam, P.; Priya, P., Saraswathyamma, C.K. & Thulaseedharan, A. (2004). Identification, Cloning and Sequence Analysis of a Dwarf Genome-Specific RAPD Marker in Rubber Tree *Hevea brasiliensis* Muell. Arg. *Plant Cell Reports*, 23, 327-332

Venkatachalam, P.; Thomas, S., Priya, P., Thanseem, I., Gireesh, T., Saraswathyamma, C.K. & Thulaseedharan, A. (2002). Identification of DNA Polymorphism with The Cultivated Clones of Rubber Tree (*Hevea brasiliensis* Muell. Arg.). *Indian J. Nat. Rubber Res.*, 15, 72-181, ISSN: 0970-2431

Waugh, R.; Baird, E. & Powell, W. (1992). The Use of RAPD Markers for The Detection of Gene Introgression in Potato. *Plant Cell Rep.*, 11, 466-469

Wendel, J.F. & Doyle, J.J. (1998). Phylogenetic Incongruence: Window into The Genome History and Molecular Evolution. In: *Molecular Systematics of Plants II. DNA Sequenceing*, D.E. Soltis, P.S. Soltis & J.J. Doyle (Ed.), 265–296, Kluwer Academic Publishing, ISBN 0-412-11121-7, Dordrecht, The Netherlands

Wendel, L.F. (1989). Visualization and Interpretation of Plant Isozymes. In: *Isozymes in Plant Biology*, E.D. Soltis & P.S. Soltis (Ed.), 5–45, Dioscorides Press, Oregon, USA

Whitkus, R.; Doebley, J., Wendel, J.F. (1994). Nuclear DNA Markers in Systematics and Evolution. In: *DNA Based Markers in Plants*, L. Phillips & I.K. Vasil (Ed.), 116–141, Kluwer Academic Publishers, ISBN 0-792-36865-7, Dordrecht, The Netherlands

Williams, J.G.K.; Kubelik, A.R., Livak, K.J., Rafalski, J.A. & Tingey, S.V. (1990). DNA Polymorphisms Amplified by Arbitrary Primers Are Useful As Genetic Markers. *Nucleic Acids Research*, 18, 6531–6535

Wycherley, P.R. (1992). The Genus *Hevea*: Botanical Aspects. In: *Natural Rubber: Biology, Cultivation and Technology*, M.R. Sethuraj & N.M. Mathew (Ed.), 50–66, Elsevier, ISBN 0-444-88329-0, Amsterdam, The Netherlands

Wycherly, P.R. (1969). Breeding of *Hevea*. *J. Rubber Res. Inst. Malaya*, 21, 38–55

3

Genetic Diversity in *Gossypium* genus

Ibrokhim Y. Abdurakhmonov et al*

Center of Genomic Technologies, Institute of Genetics and Plant Experimental Biology
Academy of Sciences of Uzbekistan, Yuqori Yuz, Qibray Region, Tashkent
Uzbekistan

1. Introduction

Cotton (*Gossypium* spp.) is the unique, most important natural fiber crop in the world that brings significant economic income, with an annual average ranging from $27 – 29 billion worldwide from lint fiber production (Campbell et al., 2010). The worldwide economic impact of the cotton industry is estimated at ~$500 billion/yr with an annual utilization of ~115-million bales or ~27-million metric tons (MT) of cotton fiber (Chen et al., 2007). In 2011 and 2012, global cotton production is projected to increase 8% (to 26.9 million MT). This will be the largest crop since 2004 and 2005 (International Cotton Advisory Committee [ICAC], 2011).

Cotton is also a significant food source for humans and livestock (Sunilkumar et al., 2006). Cotton fiber production and its export, being one of the main economic resources, annually brings an average of ~$0.9 to 1.2 billion economic income for Uzbekistan (Abdurakhmonov, 2007) that represented 22% of all Uzbek exports from 2001-2003 (Campbell et al., 2010). The economic income from cotton production accounts for roughly 11% of the Uzbekistan's GDP in 2009 (http://www.state.gov/r/pa/ei/bgn/2924.htm, verified on September 15, 2011).

The level of genetic diversity of crop species is an essential element of sustainable crop production in agriculture, including cotton. The amplitude of genetic diversity of *Gossypium* species is exclusively wide, encompassing wide geographic and ecological niches. It is conserved *in situ* at centers for cotton origin (Ulloa et al., 2006) and preserved *ex situ* within worldwide cotton germplasm collections and materials of breeding programs. Cotton

*Zabardast T. Buriev[1], Shukhrat E. Shermatov[1], Alisher A. Abdullaev[1], Khurshid Urmonov[1], Fakhriddin Kushanov[1], Sharof S. Egamberdiev[1], Umid Shapulatov[1], Abdusttor Abdukarimov[1], Sukumar Saha[2], Johnnie N. Jenkins[2], Russell J. Kohel[3], John Z. Yu[3], Alan E. Pepper[4], Siva P. Kumpatla[5] and Mauricio Ulloa[6]

[1]*Center of Genomic Technologies, Institute of Genetics and Plant Experimental Biology, Academy of Sciences of Uzbekistan, Yuqori Yuz, Qibray Region, Tashkent, Uzbekistan*
[2]*United States Department of Agriculture-Agriculture Research Service, Crop Science Research Laboratory, Starkville, Mississippi State, USA*
[3]*United States Department of Agriculture-Agriculture Research Service, Crop Germplasm Research Unit, College Station, Texas, USA*
[4]*Department of Biology, Texas A&M University, College Station, Texas, USA*
[5]*Department of Biotechnology Regulatory Sciences, Dow AgroSciences LLC, Indianapolis, Indiana, USA*
[6]*United States Department of Agriculture-Agriculture Research Service, Western Integrated Cropping Systems, USA*

productivity and the future of cotton breeding efforts tightly depend on 1) the level of the genetic diversity of cotton gene pools and 2) its effective exploitation in cotton breeding programs. Elucidating the details of genetic diversity is also very important to determine timeframe of cotton agronomy, develop a strategy for genetic gains in breeding, and conserve existing gene pools of cotton.

During past decades, because of advances in molecular marker technology, there have been extensive efforts to explore the molecular genetic diversity levels in various cotton gene pools and genomic groups, varietal and breeding collections, and specific germplasm resources. These efforts reinforced a serious concern about the narrow genetic base of cultivated cotton germplasm, which has obviously been associated with a "genetic bottleneck" occurred during historic cotton domestication process (Iqbal et al., 2001). A narrow genetic base of cultivated germplasm was one of the major factors causing the recent cotton yield and quality declines (Esbroeck & Bowman, 1998; Paterson et al., 2004).These declines, however, are largely due to challenges and the lack of innovative tools to effectively exploit genetic diversity of *Gossypium* species. The most effective utilization of genetic diversity of *Gossypium* species further requires modern genomics technologies that help to reveal the molecular basis of genetic variations of agronomic importance. Sequencing the cotton genome(s) (Chen et al., 2007;) is a pivotal step that will facilitate the fine-scale mapping and better utilization of functionally significant variations in cotton gene pools (Abdurakhmonov, 2007). Once exploited effectively, these wide ranges of genetic diversity of the genus, in particular reservoir of potentially underutilized genetic diversity in exotic wild cotton germplasm, are the 'golden' resources to improve cotton cultivars and solve many fundamental problems associated with fiber quality, resistance to insects and pathogens and tolerance to abiotic stresses (Abdurakhmonov, 2007). In this chapter, we describe cotton germplasm resources, the amplitude of morphobiological and agronomic diversity of *Gossypium* genus and review efforts on molecular genetic diversity of cotton gene pools as well as highlight examples, challenges and perspectives of exploiting genetic diversity in cotton.

2. Description of cotton gene pools and worldwide germplasm collections

Although wild cottons (*Gossypium* spp) are perennial shrubs and trees, the domesticated cottons are tropic and sub-tropic annual crops cultivated since prehistoric times of the development of human civilization. The *Gossypium* genus of the *Malvaceae* family contains more than 45 diploid species and 5 allotetraploid species (Fryxell et al., 1992; Percival et al., 1999; Ulloa et al., 2007). These species are grouped into nine genomic types ($x = 2n = 26$, or $n = 13$) with designations: AD, A, B, C, D, E, F, G, and K (Percival et al., 1999). The species are largely spread throughout the diverse geographic regions of the world. Based on the usage of these *Gossypium* species in cotton breeding and their genetic hybridization properties, they can be grouped into 1) primary gene pool, which includes the two species from the New World, *G. hirsutum* L. and *G. barbadense* L., as well as remaining three wild tetraploid species, *G. tomentosum* Nuttall *ex* Seemann, *G. mustelinum* Miers *ex* Watt and *G. darwinii* Watt; 2) secondary gene pool, including A, B, D and F genome diploid cotton species; 3) tertiary gene pool, including C, E, G, K genome *Gossypium* species (Stelly et al., 2007; Campbell et al., 2010).

Diploid cottons, referred as Old World cottons, are classified into eight (A-G to K) cytogenetically defined genome groups that have African/Asian, American, and Australian origin (Endrizzi et al., 1985). Two of these Old World cottons from Asian origin, *G. arboreum* L. and *G. herbaceum* L., with a spinnable seed fiber, were originally cultivated in Asian continent. Today, Old World cultivated cottons remain primarily for non-industrial consumption in India and adjacent Asian countries.

The New World diploid *Gossypium* comprises of 14 (one undescribed taxon US-72) D genome species (Ulloa et al., 2006; Alvarez and Wendel, 2006; Feng et al., 2011). Taxonomically, these species are recognized as the *Houzingenia* subgenus (Fryxell, 1979, 1992). Twelve of the 14 species of this group are distributed in Mexico and extending northward into Arizona. Five species are adapted to the desert environments of Baja California [*G. armourianum* Kearney (D_{2-1}), *G. harknessii* Brandegee (D_{2-2}), and *G. davidsonii* Kellogg (D_{3-d})] and NW mainland Mexico [*G. turneri* Fryxell (D_{10}) and *G. thurberi* Todaro (D_1)]. An additional seven species [*G.* sp. US-72, *G. aridum* (Rose & Standley) Skovsted (D_4), *G. lobatum* Gentry (D_7), *G. laxum* Phillips (D_9), *G. schwendimanii* Fryx. & Koch (D_{11}), *G. gossypioides* (Ulbrich) Standley (D_6), and *G. trilobum* (Mociño & Sessé ex DC.) Skovsted (D_8)] are located in the Pacific coast states of Mexico and, with the exception of the last species, are arborescent in growth habit (Ulloa et al., 2006). The other two species with disjunct distributions, *G. raimondii* Ulbrich (D_5) is endemic to Peru, while *G. klotzschianum* Andersson (D_{3-k}) is found in the Galápagos Islands. The D-genome species (subgenus *Houzingenia*) are classified into six sections: Section *Houzingenia* Fryxell (D_1 and D_8); Section *Integrifolia* Todaro (D_{3-d} and D_{3-k}); Section *Caducibracteolata* Mauer (D_{2-1}, D_{2-2}, and D_{10}); Section *Erioxylum* Rose & Standley (US-72, D_4, D_7, D_9, and D_{11}); Section *Selera* (Ulbrich) Fryxell (D_6); and Section *Austroamericana* Fryxell (D_5) (Percival et al., 1999).

Until recently, evaluation of the New World D-genome species of *Gossypium*, especially Section *Houzingenia* and Section *Erioxylum*, has been limited by the lack of resource material for *ex situ* evaluation. In recent years, the United States Department of Agriculture and the Mexican Instituto Nacional de Investigaciónes Forestales Agricolas y Pecuarias (INIFAP) have sponsored joint *Gossypium* germplasm collection trips by U.S. and Mexican cotton scientists (Ulloa et al., 2006; Feng et al., 2011). As a result of these efforts, a significant number of additional *Gossypium* accessions of the subgenus *Houzingenia* from various parts of Mexico are now available for evaluation, including several accessions of each of the arborescent species (Ulloa et al., 2006). Although none of these diploid species produces cotton fibers, the D genome is one of the parental lineages of the modern allotetraploid cultivated cottons, Upland and Pima (Ulloa, 2009). Studying these D genome species is the first critical step to fulfill the pressing need to document the *in situ* conservation, to assess the genetic diversity in *Gossypium* species for the preservation of the D genome species, and to facilitate their use for cotton improvement. *In situ* conservation of some of these species is threatened by population growth and industrialized agriculture. These *Gossypium* species are donors of important genes for cotton improvement (Ulloa et al., 2006).

Hybridization between A-genome (Old World cottons) and D-genome (New World cottons) diploids and subsequent polyploidization about 1.5 million years ago created the five AD allotetraploid lineages belonging to the primary gene pool that are indigenous to America and Hawaii (Phillips, 1964; Wendel & Albert, 1992; Adams et al., 2004). These New World allotetraploid cottons include the commercially important species, *G. hirsutum* and *G.*

barbadense, which are extensively cultivated worldwide (Abdurakhmonov, 2007; Campbell et al., 2010).

G. hirsutum (also called Acala or Upland, short stapled, Mocó, and Cambodia cotton) is the most widely cultivated (90%) and industrial cotton among all *Gossypium* species. It includes the Upland cotton cultivars and other early maturing, annually grown herbal bushes. The center of origin for *G. hirsutum* is Mesoamerica (Mexico and Guatemala), but it spread throughout Central America and Caribbean. According to archaeobotanical findings, *G. hirsutum* probably was domesticated originally within the Southern end of Mesoamerican gene pool (Wendel, 1995; Brubaker et al., 1999). Consequently, two centers of genetic diversity exist within *G. hirsutum*: Southern Mexico-Guatemala and Caribbean (Brubaker et al., 1999); Mexico-Guatemala gene pool is considered the site of original domestication and primary center of diversity. Within this range, *G. hirsutum* exhibits diverse types of morphological forms, including wild, primitive to domesticated accessions. According to Mauer (1954), there are four groups of sub-species of *G. hirsutum*: (1) *G. hirsutum* ssp. mexicanum, (2) *G. hirsutum* ssp. paniculatum, (3) *G. hirsutum* ssp. punctatum, and (4) *G. hirsutum* ssp. euhirsutum (domesticated cultivars). These four groups of sub-species include within themselves a number of wild landraces and primitive predomesticated forms such as yucatanense, richmondi, punctatum, latifolium, palmeri, morilli, purpurascens and their accessions as well as a number of domesticated variety accessions from 80 different cotton growing countries worldwide (Sunilkumar et al., 2006; Lacape et al., 2007; Abdurakhmonov, 2007).

G. barbadense (also called as long staple fibered Pima, Sea Island or Egyptian cotton), accounting for about 9% of world cotton production, was originally cultivated in coastal islands and lowland of the USA and became known as Sea Island cotton. Sea Island cottons, then, were introduced into Nile Valley of Egypt and widely grown as Egyptian cotton to produce long staple fine fibers (Abdalla et al., 2001). The wide-distribution of *G. barbadense* included mostly South America, southern Mesoamerica and the Caribbean basin (Fryxell, 1979). *G. barbadense* can be divided into two botanical races *brasilense* (with kidney-seed trait) and *barbadense* (with nonaggregated seeds) that both widely present as semi-domesticated forms in Brazil (de Almeida et al., 2009). The *brasilense* race, considered to have been domesticated in the Amazonian basin (de Almeida et al., 2009) is considered a locally domesticated form for *G. barbadense* cotton (Brubaker et al., 1999; de Almeida et al., 2009).

The other three AD tetraploid species of cotton, *G. mustelinum* with specific distribution in the Northeast Brazil (Wendel et al., 1994), *G. darwinii* endemic to Galapagos Islands (Wendel & Percy, 1990), and *G. tomentosum* Nutall ex Seemann endemic to Hawaiian Islands (DeJoode and Wendel, 1992; Hawkins et al., 2005), are truly wild species (Westengen et al., 2005).

The main *ex situ* cotton germplasm collections are in the US, France, China, India, Russia, Uzbekistan, Brazil, and Australia. Although there are a few other cotton germplasm collections present in other countries of the world, these eight countries represent the majority of the world's cotton germplasm resources. Each country has a germplasm storage and conservation program in place (Campbell et al., 2010). The history of collecting an initial cotton germplasm through the specific expeditions of cotton scientists to the centers of *Gossypium* origins are well described by Ulloa et al. (2006) that were the basis, perhaps, for the majority of the current cotton germplasm collections worldwide. Consequently, to protect the world-wide economic value of cotton and cotton byproducts, cotton germplasm collections worldwide were enriched with numerous cotton germplasm accessions and breeding materials/lines as source of the genetic diversity through continuous research efforts of specific cotton breeding programs and mutual germplasm exchange over the last 100 years (Abdurakhmonov, 2007; Campbell et al., 2010).

The brief descriptions for some of worldwide cotton germplasm collections were highlighted in several documents by Abdurakhmonov (2007), Chen et al. (2007), Stelly et al. (2007), Ibragimov et al. (2008), Wallace et al. (2009) and Campbell et al. (2010). In particular, a recent report of cotton researchers published in Crop science journal (Campbell et al., 2010) has widely described the current status of global cotton germplasm resources. Campbell et al. (2010) provided information regarding: 1) members of the collection, 2) maintenance and storage procedures, 3) seed request and disbursement, 4) funding apparatus and staffing, 5) characterization methodology, 6) data management, and 7) past and present explorations.

The contents and distribution of cotton germplasm accessions across the eight collections is summarized by Campbell et al. (2010), so we will not review the details of each collection to avoid redundancy, but rather found appropriate to list brief information in regards to the overall content and specificity of these world cotton collections. Based on a number of preserved cotton accessions in the collection, the eight major world collections can be positioned as follows: Uzbekistan (18971 accessions), India (10469 accessions), USA (10318 accessions), China (8837 accessions), Russia (6276 accessions), Brazil (4296 accessions), CIRAD (France; 3070 accessions) and Australia (1711 accessions). The main content of these collections consists of accessions for two cultivated cotton species, G. hirsutum and G. barbadense. Uzbekistan (2680 accessions), India (2283 accessions) and USA (1923 accessions) collections are the richest ones to maintain a great number of accessions for Asian diploid cottons, G. herbaceum and G. arboreum belonging to the secondary gene pool. If the collection of wild species belonging to primary, secondary and tertiary gene pools are considered Brazil (889 accessions), USA (509 accessions) and CIRAD (295 accessions) are the richest cotton collections in the world.

3. Spectra of morphological and agronomic diversity in cotton

The amplitude of genetic diversity of cotton (Gossypium spp), including all its morphological, physiological and agronomic properties, is exclusively wide (Mauer, 1954). There is a great deal of genetic diversity in the Gossypium genus with characteristics such as plant architecture, stem pubescence and color, leaf plate shape, flower color, pollen color, boll shape, fiber quality, yield potential, early maturity, photoperiod dependency, and resistance to multi-adversity environmental stresses that are important for the applied breeding of cotton. The glimpse of genetic diversity on some morphological traits is demonstrated in Figs. 1 and 2.

Besides morphological diversity in Gossypium genus, representatives of different genomic groups have diverse characteristics in many agronomically useful traits. Considering only G. hirsutum accessions, exotic and cultivar germplasm represent a wide range of genetic diversity in yield and fiber quality parameters. For example, in the analyses of ~1000 G. hirsutum exotic and cultivated accessions in the two different environments, Mexico and Uzbekistan, we found a wide range of useful agronomic diversities (Abdurakhmonov et al., 2004, 2006, 2008, 2009). In one or two environments, the cotton boll mass varies in a range of

1-9 grams per boll, 1000 seed mass varies in a range of 50-170 grams, the lint percentage varies in a range of 0-45%, Micronaire varies in a range of 3-7 mic, the fiber length varies in a range of 1-1.28 inch, and fiber strength varies in a range of 26-36 g/tex. There was also a wide range of variation in photoperiodic flowering (day neutral, weak to strong photoperiodic dependency) and maturity (Abdurakhmonov, 2007). This wide phenotypic diversity of cotton shows the extensive plasticity of cotton plants and potential of their wide utilization in the breeding programs as an initial material.

4. Some examples of exploiting genetic diversity through traditional breeding

Above mentioned genetic diversity, preserved in germplasm collections worldwide, are the golden resources to genetically improve the cotton cultivars. There are numerous examples on the utilization of such genetic variations in solving many fundamental problems in cotton breeding and production (Abdurakhmonov, 2007). For instance, the exploration for genetic diversity for *Verticillium* wilt fungi from the exotic *G. hirsutum ssp mexicanum var nervosum* germplasm and its on-time mobilization into the elite cultivars solved wilt epidemics in 1960's and saved Uzbekistan's cotton production, and so the economy of the country (Abdullaev et al., 2009). As a result, the wilt resistant variety series named as "Tashkent" were developed (Abdukarimov et al., 2003; Abdurakhmonov, 2007). Later, salt tolerant genotype AN-Boyovut-2 was selected from Tashkent cultivar biotypes demonstrating a continuation of a 'genetic diversity imprint' introgressed from the wild landrace stock (Abdukarimov et al., 2003). This is one of the success stories on exploiting genetic diversity and its impact from the single landrace stock germplasm, *G. hirsutum* ssp. mexicanum (Abdurakhmonov, 2007). A number of other examples on the creation of natural defoliation, disease and pest resistance, tolerance to multi-adversity stresses, improved seed oil content and fiber quality parameters, utilizing the exotic germplasm genetic diversity in Uzbekistan have been well documented (Abdukarimov et al., 2003; Abdurakhmonov et al., 2005, 2007).

Successful photoperiodic conversion program in cotton was developed to mobilize day-neutral genes into the primitive accessions of *G. hirsutum*. Day-neutral genes were introgressed into 97 primitive cotton accessions by a large backcrossing effort (McCarty et al., 1979; McCarty & Jenkins, 1993, Liu et al., 2000). This converted cotton germplasm is an important reservoir for potential genetic diversity and can be used as a source to introgress genes into breeding germplasm (Abdurakhmonov, 2007).

Similarly, using genetic diversity existing in *Gossypium* genus, reniform nematode resistance, which is one of the high cost ($100 million/year) problems in US cotton production, was addressed. Scientists succeeded in introgressing high resistance to the nematode from G. *longicalyx* into *G. hirsutum* through genetic bridge crossing of two trispecies hybrids of G. *hirsutum*, *G. longicalyx*, and either *G. armourianum* or *G. herbaceum* (Robinson et al., 2007). Later, a gene of interest was mapped (Dighe et al., 2010). Resistance to root-knot nematode was also solved with the use of genetic diversity in *Gossypium* genomes (Roberts & Ulloa, 2010). Additionally, Hinze et al. (2011) developed four diverse populations based on US germplasm collection that helped to utilize a large amount of 'still underutilized' genetic variability in cotton breeding that should be useful in sustainable cotton production with superior quality. There are many other examples recorded in different cotton breeding programs, but we limit this section with above examples and move to address the challenges behind these success stories and future perspectives in this direction.

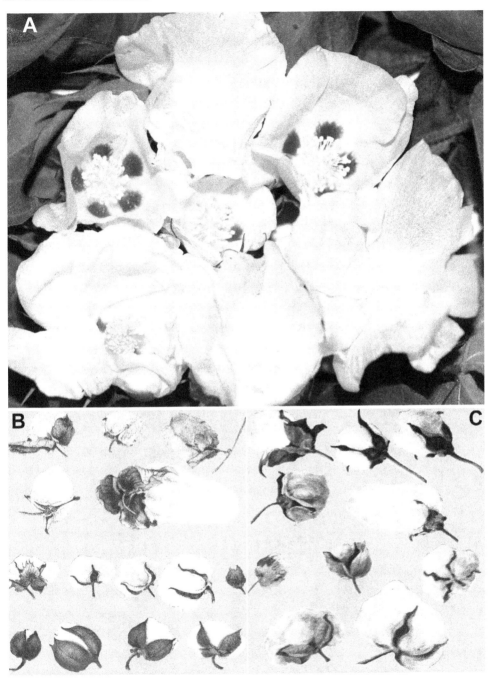

Fig. 1. Morphological trait diversity: (A)-pollen color, petal color and spot; (B)-matured bolls in diploid species, and (C)-matured bolls in tetraploid species.

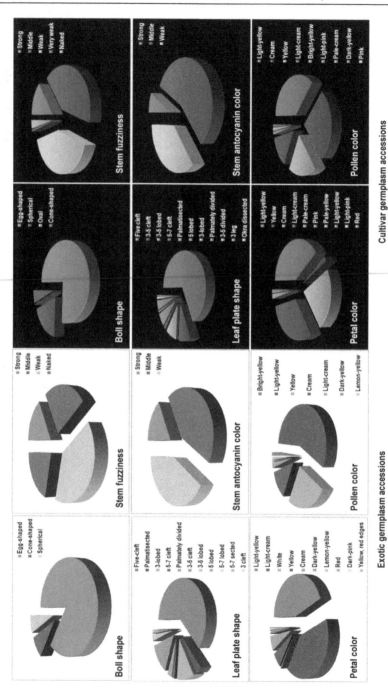

Fig. 2. Diversity on several morphological traits in exotic and cultivar germplasm of *G. hirsutum* accessions from Uzbekistan cotton collection (Abdurakhmonov et al., 2004, 2006).

5. Challenges and perspectives of exploiting diversity of different gene pools

The introduction of genetic diversity into elite cotton germplasm is difficult and the breeding process is slow. When breeders use new and exotic germplasm sources, which possess desirable genes for crop trait improvements, large blocks of undesirable genes are also introgressed during the recombination between the two parental lines (linkage drag). This linkage drag has limited the use of such germplasm. Therefore, the utilization of useful genetic diversity of the wild germplasm using traditional breeding efforts is challenging due to: 1) hybridization issues between various cotton genomes, 2) sterility issues of interspecific multi-genome hybrids, 3) segregation distortion, 4) photoperiodic flowering of wild cottons and 5) long timescale (10-12 years of efforts) required for successful introgression and recovering superior quality homozygous genotypes using traditional breeding approaches (Abdurakhmonov, 2007). This underlies necessity for the development of new innovative genomics approaches to support and accelerate the traditional efforts of exploiting the genetic diversity in cotton breeding. Continuing the introduction of genetic diversity into cultivated plants is important for reducing crop vulnerability and improving important traits such as yield, fiber quality traits, and disease and pest resistance of the cotton crop.

The most effective utilization of the genetic diversity of *Gossypium* species further requires (1) characterization of candidate gene(s) underlying the phenotypic and agronomic diversities based on genomic information in other species, (2) estimation of molecular diversity, genetic distances, genealogy and phylogeny of gene pools and germplasm groups, (3) acceleration of linkage mapping and marker-assisted selection, (4) development of efficient cotton transgenomics, and (5) sequencing cotton genome(s) (Abdurakhmonov, 2007). Furthermore, (6) it is very important to characterize and describe the existing cotton germplasm collections for both phenotypic and genomic diversity. Consequently, (7) incorporation of information into electronic web-based cotton databases such as cotton DB (http://cottondb.org), Cotton Portal (http://gossypium.info), and the Cotton Diversity Database (http://cotton.agtec.uga.edu; Gingle et al., 2006) as well as further improvement of data management tools are pivotal to facilitate an effective exploitation of the genetic diversity of cotton in the future. Cotton germplasm exchange (8) among collections and research groups is also an imperative part toward this goal (Abdurakhmonov, 2007).

6. Characterization of molecular genetic diversity in *Gossypium* genus

Molecular diversity using protein and DNA marker technologies has extensively been studied for accessions from primary and secondary gene pools. Molecular genetic diversity of tertiary gene pool cotton species is poorly explored using molecular marker technology .

6.1 Molecular diversity within primary gene pool

6.1.1 Upland germplasm

As mentioned above, application of modern molecular marker technologies, such as isozymes (Wendel & Percy, 1990; Wendel et al., 1992), random amplified polymorphic DNAs – RAPDs (Multani & Lyyon, 1995; Tatineni et al., 1996; Iqbal et al., 1997; Mahmood et al., 2009; Chaudhary et al., 2010), restricted fragment length polymorphisms – RFLPs (Wendel & Brubaker, 1993), amplified fragment length polymorphisms – AFLPs (Pilay & Myers, 1999; Abdalla et al., 2001; Iqbal et al., 2001; Rana et al., 2005; Lukonge et al., 2007) and

Simple Sequence Repeats – SSRs (Liu et al., 2000, Gutierrez et al., 2002; Rungis et al., 2005; Zhang et al., 2005a; Bertini et al., 2006; Zhang et al., 2011a; Kalivas et al., 2011) generally revealed a low level of genetic diversity within Upland cultivars. There were little variations in estimation of molecular diversity among Upland cultivars (G. hirsutum); however, in general, the genetic distance reported for Upland cultivars was in the range of 0.01-0.28 (Abdurakhmonov, 2007).

Recently, we analyzed a large number of G. hirsutum variety and exotic accessions from Uzbek cotton germplasm collection (Fig.2) with SSR markers (Abdurakhmonov et al., 2008, 2009). Analysis of a large number of G. hirsutum accessions from exotic germplasm and diverse ecotypes/breeding programs with SSR markers confirmed the narrow genetic base of Upland cotton cultivar germplasm pool (with the genetic distance (GD) range of 0.005-0.26) and provided an additional evidence for the occurrence of a genetic 'bottleneck' during domestication events of the Upland cultivars at molecular level (Iqbal et al., 2001). Molecular diversity analysis of germplasm accessions using principal component analysis (PCA) suggested that germplasm resources could be broadly grouped into three large clusters (Fig.3) of exotic (1), USA-type (2) and Uzbekistan (3). First three eigenvalues of PCA analysis accounted for a ~52% variation and demonstrated existence of wide genetic diversity within the exotic germplasm, including germplasm accessions from Mexican and African origin (GD=0.02-0.50; Fig.3). We recorded a plenty of private SSR alleles within each group of accessions, specific to the germplasm groups, breeding ecotypes or exotic accessions.

A wider genetic diversity in the land race stocks of G. hirsutum was reported by previous studies (Liu et al., 2000, Lacape et al., 2007), suggesting the existence of sufficient genetic diversity in the exotic germplasm for future breeding programs. Rana et al. (2005) also reported a wider genetic diversity (30-87%) within G. hirsutum breeding lines using AFLP markers. Some recent studies have reported a relatively higher genetic diversity with an average genetic distance of up to ~37-77% in G. hirsutum cultivars, based on the analysis of specific germplasm resources from Brazil (Bertini et al., 2006), Pakistan (Khan et al., 2009; Azamat & Khan, 2010), India (Chaudhary et al., 2010) and China (Liu et al., 2011; Zhang et al., 2011a) breeding programs. Results of these studies were inferred from SSR or combination of a SSR and/or RAPD marker polymorphisms.

Similarly, using SSR and RAPD markers, Sapkal et al. (2011) reported moderately high level of genetic diversity (up to 57%) for 91 Upland cotton accessions with genetic male sterility maintainer and restorer properties. This suggested the existence of useful genetic diversity both in exotic and breeding line resources, useful to broaden the genetic base of Upland cotton cultivars. There is a need for evaluation of molecular genetic diversity level (Zhang et al., 2011a) and its effective exploitation in breeding programs that will address current concerns on narrowness of genetic base of widely grown Upland cotton cultivars (Hinze et al., 2011).

6.1.2 Sea Island germplasm

The molecular genetic diversity within G. barbadense germplasm accessions was also studied using molecular markers such as allozymes (Wendel & Percy, 1990) and AFLPs (Abdalla et al., 2001; Westengen et al., 2005). These studies revealed a narrow genetic base within G. barbadense accessions with a genetic distance of 7-11% (Abdalla et al., 2001; Westengen et al., 2005)

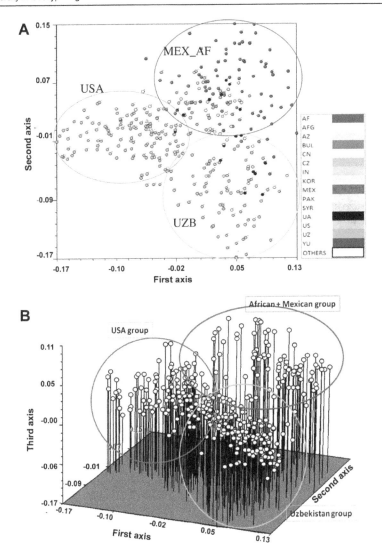

Fig. 3. Principal coordinate analysis of Upland cotton (*G. hirsutum*) accessions from Uzbek cotton germplasm collection analyzed with SSR markers. Two (A) and three (B) dimensional view for accessions from Africa (AF), Afghanistan (AFG), Bulgaria (BUL), China (CN), Czechoslovakia (CZ), India (IN), Korea (KOR), Mexico (MEX), Pakistan (PAK), Syria (SYR), Ukraine (UA), United States (US), Uzbekistan (UZB), Yugoslavia (YU), and others (Turkey, Iraq and Azerbaijan).

as was observed within the Upland cotton germplasm. In contrast, Boopathi et al. (2008) have identified highly diverse pairs of *G. barbadense* accessions using SSR marker analysis, which is useful for breeding of high quality Pima type cotton cultivars. Recently, de Almeida et al. (2009) have studied the molecular diversity level of *G. barbadense* populations *in situ* preserved in the two states of Brazil, Ampa and Para. The genetic analysis using SSR markers of plant populations in these two states revealed 1) high homozygosity in each genotype tested, 2) high total genetic diversity (H_e=39%) in *G. barbadense* populations studied and 3) high level of population differentiation (F_{st}=36%) between cotton plants from these two Brazilian states. Results suggested the existence of noticeable genetic diversity preserved in *in situ* populations of *G. barbadense* in Brazil that should be further maintained within an *ex situ* germplasm collection to guarantee its long term preservation (de Almeida et al., 2009). Similarly, there is useful genetic diversity in *ex situ* preserved *G. barbadense* germplasm collections worldwide. For instance, the molecular diversity analysis of *G. barbadense* accessions using SSR markers revealed that moderately higher genetic diversity (up to 34%) exists within former USSR (that includes collections of Uzbekistan and Russia), China, USA, and Egypt germplasm collections (Wu et al., 2010). In that, USSR collection demonstrated the extraordinary genetic diversity compared with other collections whereas Egyptian collection had the least genetic diversity.

6.1.3 Wild allotetraploid germplasm

The molecular diversity revealed by AFLP markers was low within *G. tomentosum* germplasm with a genetic distance range of 2-11% (Hawkins et al., 2005). However, recent efforts on the characterization of genetic diversity level of three *in situ* preserved *G. mustelinum* population from Brazil using SSR markers suggested 1) high level of homozygosity within each population studied and 2) existence of high level of total genetic differentiations (58.5%) between them, which is due to geographic isolations and genetic founder effects (Barrosso et al., 2010). Wendel & Percy (1990) analyzed 58 *G. darwinii* accessions from six islands using 17 isozyme markers encoded by 59 genetic loci and identified high genetic diversity level within its accessions and relationships with *G. barbadense* and *G. hirsutum* genomes. This classical study suggested that *G. darwinii* is closely related to *G. barbadense* despite having gene flow imprints from *G. hirsutum;*   however, *G. darwinii* has a large number of unique alleles to be considered a distinct genome (Wendel & Percy, 1990).

6.2 Molecular diversity within secondary gene pool

The genomic diversity of the A-genome diploid cottons has also been studied using molecular marker technology (Liu et al., 2006; Guo et al., 2006; Kebede et al., 2007; Rahman et al., 2008; Kantartzi et al., 2009; Patel et al., 2009; Azamat & Khan, 2010). The genetic distance within 39 *G. arboreum* L (A_2A_2-genome) accessions, analyzed with SSR markers, ranged from 0.13-0.42 (Liu et al., 2006) demonstrating the existence of wider genomic diversity in the A-genome diploids compared to the Upland cultivar germplasm. Kebede et al. (2007) reported, however, moderate level of genetic diversity within each A_1 and A_2-genome cottons that ranged from 0.03-0.20 with an average of 0.11 within *G. herbaceum* and 0.02-0.18 with an average of 0.11 for *G. arboreum* (A_2). The overall genetic distance between A_1 and A_2 genomes was up to 36-38% (Kebede et al., 2007; Mahmood et al., 2010). In fact, *G. arboreum* arose from the primitive perennial form of *G. herbaceum* spread in India and there is a single reciprocal chromosomal translocation in *G. arboreum* genome compared to *G. herbaceum* (Guo et al., 2006). Molecular diversity revealed by SSR markers was higher within *G. arboreum* accessions (an average of 25%) compared to *G. herbaceum* accessions (an average of 4%; Patel et al., 2009), suggesting differences in two closely related cotton genome germplasm resources. This is an interesting finding but is in contrast to the report by Kebede et al. (2007) where an average genetic diversity within A_1 and A_2 genome accessions was equal.

Rahman et al. (2008) studied 32 G. *arboreum* accessions specific to Pakistan with RAPD markers and found up to 53% genetic diversity between studied accessions with very narrow diversity within cultivated G. *arboreum* accessions compared to non-cultivated ones. Analyzing 96 G. *arboreum* accessions with SSR markers, Kantartzi et al. (2009) reported that genetic distance within these geographically diverse A_2 genome accessions ranged up to 51%. In a more recent study, Azamat & Khan (2010) also reported wider genetic diversity in G. *arboreum* cultivar germplasm revealed by RAPD (GD=0.371) and SSR markers (GD=0.41). Although variable genetic distance estimates are presented, these reports collectively suggest that A genome representatives of secondary gene pool have sufficient molecular diversity useful for breeding programs.

Studying a large number of accessions for D genome cotton such as G. *aridum* (D_4), G. *davidsonii* (D_{3-d}), G. *klotzschianum* (D_{3-k}), G. laxum (D_9), G. lobatum (D_7), G. *schwendimanii* (D_{11}) with AFLP markers Alvarez & Wendel (2006) have reported 7 to 54% genetic diversity among D-genome accessions studied. A wider range of genetic diversity was observed among 12 D-genome diploid cottons with the genetic similarity of 0.08-0.94 (Guo et al., 2007a), suggesting existence of diverse variations in D-genome cotton germplasm useful for breeding programs. Recently, Feng et al. (2011) have studied 33 arborescent D-genome accessions, including 23 accessions of G. *aridum* with RAPD and AFLP markers. They found high molecular diversity among accessions studied, varying from 32% to 84%. This study suggests for continual efforts to study these D-genome American *Gossypium* species (subsection *Erioxylum*) to resolve genetically distant geographical ecotypes useful for cotton improvement (Feng et al., 2011)

6.3 Molecular diversity within tertiary gene pool

There is a limited information on molecular diversity estimates for tertiary germplasm pool accessions, including C, E, G and K-genomic species. Recently, Tiwari & Stewart (2008) reported AFLP marker-based molecular diversity analysis results for 57 accessions of C- and G-genome species, including G. *australe* F. Mueller *(G)*, G. *nelsonii* Fryxell (G_3), G. *bickii* Prokhanov (G_1) and G. *sturtianum* J.H. Willis (C_1). Results showed that within G. *australe* accessions, the pairwise mean genetic distance was in a range of 3-15%, suggesting narrow genetic diversity within G. *australe* accessions that could be due to relatively recent seed dispersal over large growing area of this species (Tiwari & Stewart, 2008). However, there was moderately high molecular diversity between G. *australe* and G. *nelsonii* accessions, ranging from ~17-31%.Higher molecular diversity of up to ~43% was found between G. *australe* and G. *bickii* accessions. The genetic distance between G. *bickii* and G. *nelsonii* varied from 25% to 35% and as expected C_1-genome accessions were most distantly related ones to these three G-genome species (Tiwari & Stewart, 2008). There is no report on molecular diversity studies on other representatives of tertiary gene pool.

6.4 Molecular diversity among cotton gene pools

The genetic diversity among different gene pools was also estimated in many studies using various marker systems. AFLP marker analyses studies (Iqbal et al., 2001, Abdalla et al.,

2001, Westengen et al., 2005) revealed that the genetic distance between G. *barbadense* and G. *hirsutum* was in the range of 21-33%. The other wild AD tetraploids (G. *mustelinum*, G. *tomentosum*) were close to the cultivated AD cottons sharing 75-84% similarity, where G. *tomentosum* was closer to G. *hirsutum* genome (GD=0.16) than the other allotetraploid species (Westengen et al., 2005). At the same time, as mentioned above, G. *darwinii* was closer to G. *barbadense* than G. *hirsutum* (Wendel & Percy, 1990).

Based on AFLP marker analysis, the genetic distance between the widely cultivated AD cottons (G. *barbadense* and G. *hirsutum*) and A-genome diploids varied from 45 to 69%, and that between the cultivated AD cottons and the D-genome varied from 55 to 71%. The genetic distance between the wild AD tetraploids and the A-genome was in the range of 46-52%, and between the wild AD cottons and the D-genome was 58-59%. The genetic distance between the A- and D-genome cottons was in the range of 0.72-0.82 when analyzed with AFLPs (Iqbal et al., 2001, Abdalla et al., 2001, Westengen et al., 2005).

The use of SSR markers revealed that the genetic distance between G. *hirsutum* and G. *barbadense* was in a range of 42-54% (Kebede et al., 2007). However, Lacape et al. (2007) reported higher genome dissimilarity values (D=0.89-0.91%) between G. *hirsutum* and G. *barbadense* within their material. Also, high mean dissimilarity values were reported between G. *hirsutum* and G. *tomentosum* (D=0.71-0.75) and between G. *barbadense* and G. *tomentosum* (D=0.80) using highly polymorphic sets of SSRs (Lacape et al., 2007). The genetic distance among the AD tetraploids was also in the range of 0.80-0.88 (Liu et al., 2000) with moderate closeness of G. *tomentosum* to the Upland cotton than G. *barbadense* cultivars that was also supported by other studies with different marker systems (Dejoode & Wendel, 1992; Hawkins et al., 2005). Based on SSR marker analysis, the genetic distance between the cultivated AD cottons and the A-genome was in the range of 31-43%, and that between the cultivated AD cottons and the D-genome was in the range of 35-46% (Kebede et al., 2007). The genetic distance between A-and D-genome cottons varied in the range of 29-42% (Kebede et al., 2007).

7. Perspectives of 21st century cotton genomics efforts in characterizing and exploiting the genetic diversity of *Gossypium* species

During the past two decades, the international cotton research community has made extensive efforts to utilize the genetic diversity in cotton, which are imperative for the future of trait improvements of the cotton crop. There are many marker systems such as isozymes, RAPDs, RFLPs, AFLPs (extensively referenced herein), and their various modifications (Zhang et al., 2005b) successfully used in cotton. However, the development of a large collection of robust, portable, and PCR-based molecular marker resources such as Simple Sequence Repeats (SSRs; www.cottonmarker.org) and Single Nucleotide Polymorphisms (SNPs) for cotton were one of the tremendous accomplishments of cotton research community (Chen et al., 2007; Van Deynze et al., 2009). This accelerated studies on genetic diversity in cotton at genomic level. Cotton marker resources were made available for cotton research community through cotton marker database (CMD) (Blenda et al., 2006) that are being extensively used to create cotton genetic linkage maps and to map important agronomic QTLs (Abdurakhmonov, 2007; Chen et al., 2007; Zhang et al., 2008). In addition to available DNA marker systems, recently, Reddy et al. (2011) developed a diversity array technology (DArT) marker platform for the cotton genome and evaluated the use of DArT markers compared with AFLP markers in mapping populations. These studies are very important to elucidate molecular basis of genetic diversities in cotton that are vital to mobilize useful genes of agronomic importance to the elite cultivars through marker-assisted breeding programs.

Furthermore, researchers have reported several potential candidate genes of many agronomic traits in cotton. Tremendous efforts were made to study molecular basis of one of the most complex, but important traits – cotton fiber development (Abdurakhmonov, 2007; Chen et al., 2007; Zhang et al., 2008). These efforts, including many more recent reports on the dissection of candidate genes that are specifically expressed in developing fibers are undoubtedly imperative for future exploitation of genetic diversity in cotton fiber traits using transgenomics approaches (Arpat et al., 2004; Ruan et al., 2003; Zhang et al., 2011b).

Despite wide spectra of genetic diversity in *Gossypium* genus and extensive cotton genomics efforts, cotton lags behind other major crops for marker-assisted breeding due to limited polymorphism in the cultivated germplasm. This underlies broadening of cultivar germplasm genetic base through mobilization of useful gene variants from other gene pools into cultivated germplasm. There is a need for application of modern innovative genomics tools such as association mapping to identify genetic causatives of natural variations preserved in cotton germplasm resources and their use in plant breeding. Efforts on turning the gene-tagging efforts from bi-parental crosses to natural population or germplasm collections, and from now classical QTL-mapping approach to modern linkage disequilibrium (LD)-based association study should lead to elucidation of *ex situ* conserved natural genetic diversity of worldwide cotton germplasm resources and its effective utilization. LD refers to a historically reduced (non-equilibrium) level of the recombination of specific alleles at different loci controlling particular genetic variations in a population (Abdurakhmonov & Abdukarimov, 2008). Although novel to cotton research, the association genetics strategy is, in fact, highly applicable to the identification of markers linked to fiber quality and yield through the examination of linkage disequilibrium (LD) of DNA-based markers with fiber quality and yield traits in a large, diverse germplasm collection (Abdurakhmonov et al., 2004, 2008, 2009).

Application of association mapping strategy in gene mapping and germplasm characterization gained wider use in cotton. For example, Kantartzi & Stewart (2008) conducted association analysis for the main fiber traits in 56 *G. arboreum* germplasm accessions introduced from nine regions of Africa, Asia and Europe using 98 SSR markers. Association mapping strategy was also applied for tagging fiber traits in the exotic germplasm derived from multiple crosses among *Gossypium* tetraploid species (Zeng et al., 2009). Both of these studies did not quantify the LD level in the population and used marker-trait associations to tag genetic variations contributing to the trait of interest.

Alternatively, to better assess and exploit a molecular diversity of cotton genus, we conducted molecular genetic analyses in a global set of ~1000 *G. hirsutum* L. accessions, one of the widely grown allotetraploid cotton species, from Uzbek cotton germplasm collection. This global set represented at least 37 cotton growing countries and 8 breeding ecotypes as well as wild landrace stock accessions. The important fiber quality (fiber length and strength, Micronaire, uniformity, reflectance, elongation, etc.) traits were measured in two distinct environments of Uzbekistan and Mexico. This study allowed us to quantify the linkage disequilibrium level in the genome of Upland cotton germplasm and to design an "association mapping" study to find biologically meaningful marker-trait associations for important fiber quality traits that accounts for population confounding effects (Yu et al., 2006; Abdurakhmonov & Abdukarimov, 2008). Several SSR markers associated with major fiber quality traits along with donor accessions were identified and selected for MAS programs (Abdurakhmonov et al., 2008, 2009).

Further, with the specific objective of introducing and enriching the currently-applied traditional breeding approaches with more efficient modern MAS tools in Uzbekistan, we began marker-assisted selection efforts based on our association mapping studies mentioned above. For this purpose, we selected (1) a set of twenty three major (Micronaire, fiber strength and length, and elongation) fiber trait-associated DNA markers as a tool to manipulate the transfer of QTL loci during a genetic hybridization; and (2) thirty-seven (11 wild race stocks and 26 variety accessions from diverse ecotypes) donor cotton genotypes that bear important QTLs for fiber traits. These donor genotypes were crossed with 9 commercial cultivars of Uzbekistan (as recipients) in various combinations with the objective of improving one or more of fiber characteristics of these recipients. These 9 parental recipient genomes were first screened with our DNA-marker panel to compare with 37 donor genotypes. The polymorphic status of marker bands between donor and recipient genotypes were recorded. The hybrid plants generated from each crossing combination were tested using DNA-markers at the seedling stage, and hybrids bearing DNA-marker bands from donor plants were selected for further backcross breeding (Abdurakhmonov et al., 2011).

Testing the major fiber quality traits using HVI in trait-associated marker-band-bearing hybrids revealed that mobilization of the specific marker bands from donors really had positively improved the trait of interest in recipient genotypes (data not shown). Currently, we developed a second generation of recurrent parent backcrossed hybrids (F_1BC_2), bearing novel marker bands and having superior fiber quality compared to original recipient parent (lacking trait-associated SSR bands). These results showed the functionality of the trait-associated SSR markers detected in our association mapping efforts in diverse set of Upland cotton germplasm. Using these effective molecular markers as a breeding tool, we aim to pyramid major fiber quality traits into single genotype of several commercial Upland cotton cultivars of Uzbekistan. Our efforts will not only help rapid introgression of novel polymorphisms, broadening the genetic diversity of cotton cultivars and accelerating the breeding efforts for future sustainable cotton production in Uzbekistan but also exemplify effective exploitation of the natural genetic diversity *ex situ* preserved in cotton germplasm collections (Abdurakhmonov et al., 2011).

In spite of successful application of association mapping in cotton, there is a great challenge with assigning correct allelic relationships (identity by decent) of multiple band amplicons when diverse, reticulated, and polyploid cotton germplasm resources lacking historical pedigree information are investigated. Besides, there is the issue of rare and unique alleles that is problematic for conducting association mapping (Abdurakhmonov & Abdukarimov, 2008). While these issues can be solved using many available methodologies and approaches (Abdurakhmonov & Abdukarimov, 2008); however, recent studies in model crops suggested a new methodology to minimize these issues with the creation of segregating populations, performing genetic crosses between several reference populations with known allele frequencies for functional polymorphisms. Such an approach is referred to as nested association mapping (NAM) and NAM populations would greatly enhance the power of association mapping in plants (Stich and Melchinger, 2010). The usefulness and feasibility of NAM population based genetic mapping studies were successfully demonstrated in maize (Kump et al., 2011; Poland et al., 2011; Tian et al., 2011;) and should be adopted for other crops with complex genome and diverse germplasm resources like cotton. Therefore, creation of NAM populations for cotton on the basis of germplasm evaluation and characterization studies is the task of high priority for future characterization and mapping of biologically meaningful genetic variations in cotton. This requires further efforts and investments that facilitate fine-scale association mapping studies in cotton. This will ultimately lead to cloning and characterization of genetic causatives controlling the genetic diversities and its effective exploitation in plant breeding.

8. Conclusions

In conclusion, by having a wide geographic and ecological dispersal, the *Gossypium* genus represents and preserves large amplitude of morphobiological and genetic diversity within its *ex situ* worldwide germplasm collections and *in situ* occupation sites. Because of the development of molecular marker technologies, and their application in genetic diversity studies of germplasm resources, various gene pools and specific cultivar groups, researchers found a genetic bottleneck in cultivated cotton germplasm resources. However, there is moderately high molecular diversity present in some specific cultivar germplasm analyzed worldwide, suggesting a need for continual efforts on searching the diverse cultivar germplasm resources using molecular markers. There is a need to extend molecular marker-based diversity studies for tertiary gene pool accessions of cotton. There also is high genetic diversity available within exotic land race stocks, wild AD cottons, and putative A-and D-genome ancestors to AD cottons that have potential to search for genetic variations useful in future improvement of cotton. The variations observed in genetic distance estimations between different studies could be due to (1) germplasm resources chosen for the study, (2) number of accessions analyzed with molecular markers, (3) number of markers and marker types used, (4) genomic regions screened, and (5) subjective features of data analyses process, e.g., considering or removing unique or rare alleles, largely influencing the genetic distance measures.

Further, the narrowness of genetic diversity in cultivar germplasm was associated with recent and possibly future declines in cotton production and its quality, which was a timely warning to accelerate efforts on broadening the genetic base of cultivar germplasm resource via mobilizing novel genetic variants from wild, primitive, pre-domesticated primary, secondary and tertiary gene pools. Traditional efforts have succeeded in introgressing many new genetic variations into cultivar germplasm from other gene pools, but it is still challenging and the breeding process is slow due to a number of genetic barriers and obstacles, as highlighted above, to accomplish the goal. This underlies the importance of development of innovative tools to exploit the biologically meaningful genetic variations, existing in *Gossypium* genus. The most effective utilization of genetic diversity of *Gossypium* species further requires characterization of candidate gene(s) underlying the phenotypic and agronomic diversities, acceleration of linkage mapping, map-based cloning and marker-assisted selection that underlie development of modern genomics technologies such as high-resolution, cost effective LD-based association mapping for cotton with its optimization through development of modern nested association mapping populations. The development of efficient cotton transgenomics tools and complete sequencing of cotton genome(s) will further accelerate exploitation of genetic diversity in highly specific manner and with clear vision. Future application of whole genome-association strategy with epigenomics perspectives, which currently is widely being applied in human and the other model plants such as Arabidopsis, will have a significant impact on identifying true functions of genes controlling available genetic diversity, and consequently, its effective utilization.

9. Acknowledgements

We thank the Office of International Research Programs (OIRP) of United States Department of Agriculture (USDA) for continual funding of our collaborative research on cotton germplasm characterization and genetic diversity analysis. We acknowledge Civilian Research Development Foundation (CRDF), USA for project coordination and Academy of Sciences of Uzbekistan for their continual in-house support of the research efforts.

10. References

Abdalla, A.M.; Reddy, O.U.K.; El-Zik, K.M. & Pepper, A.E. (2001). Genetic Diversity and Relationships of Diploid and Tetraploid Cottons Revealed Using AFLP. *Theoretical and Applied Genetics*, Vol. 102, No. 2-3, (February 2001), pp. 222-229, ISSN 1432-2242

Abdukarimov, A.; Djataev, S. & Abdurakhmonov, I.Y. (2003). Cotton Research in Uzbekistan: Elite Varieties and Future of Cotton Breeding, *Proceedings of World Cotton Research Conference-3*, pp. 5-16, Cape Town, South Africa, March 9-13, 2003

Abdullaev, A.A.; Klyat, V.P. & Rizaeva, C.M. (2009). Cotton Introduction in Uzbekistan - History and Perspectives of Using of Plant Introduction: Problems and Perspectives. (2009), *Proceedings of 4th National scientific-applied conference*, pp. 59-61, Tashkent, Uzbekistan, July 3-4, 2009 (in Russian)

Abdurakhmonov, I.Y. (2007). Exploiting Genetic Diversity, *Proceedings of World Cotton Research Conference-4*, P2153, Lubbock, Texas, USA, September 10-14, 2007

Abdurakhmonov, I. Y. & Abdukarimov A. (2008). Application of association mapping to understanding the genetic diversity of plant germplasm resources. *International Journal of Plant Genomics*, Vol. 2008, No. 574927, (June 2008), pp.1-18, ISSN 1687-5389

Abdurakhmonov, I.Y.; Abdullaev, A.; Rizaeva, S.; Buriev, Z.; Adylova, A.; Abdukarimov, A.; Saha, S.; Kohel, R.; Yu, J. & Pepper, A.E. (2004). Evaluation of G. hirsutum Exotic Accessions From Uzbek Cotton Germplasm Collection for Further Molecular Mapping Purposes, *Proceedings Beltwide Cotton Improvement Conference*, pp. 1133-1142, San Antonio, Texas, USA, January 5-9, 2004

Abdurakhmonov, I.Y.; Abdullaev, A.A.; Saha, S.; Buriev, Z.T.; Arslanov, D.; Kuryazov, Z.; Mavlonov, G.T.; Rizaeva, S.M.; Reddy, U.K.; Jenkins, J.N.; Abdullaev, A. & Abdukarimov, A. (2005). Simple Sequence Repeat Marker Associated With a Natural Leaf Defoliation Trait in Tetraploid Cotton. *Journal of Heredity*, Vol. 96, No. 6, (September 2005), pp. 644-653, ISSN 1465-7333

Abdurakhmonov, I.Y.; Buriev, Z.T.; Salakhuddinov, I.B.; Rizaeva, S.M.; Adylova, A.T. Shermatov, S.E.; Adukarimov, A.; Kohel, R.J.; Yu, J.Z.; Pepper, A.E .; Saha, S. & Jenkins, J.N. (2006). Characterization of G. hirsutum Wild and Variety Accessions from Uzbek Cotton Germplasm Collection for Morphological and Fiber Quality Traits and Database Development, *Proceedings Beltwide Cotton Improvement Conference,* P5306, San Antonio, Texas, USA, January 3-6, 2006

Abdurakhmonov, I.Y.; Kohel, R.J.; Yu, J.Z.; Pepper, A.E.; Abdullaev, A.A.; Kushanov, F.N.; Salakhutdinov, I.B.; Buriev, Z.T.; Saha,S.; Scheffler, B.E.; Jenkins, J.N. & Abdukarimov, A. (2008). Molecular Diversity and Association Mapping of Fiber Quality Traits in Exotic *G. hirsutum* L. Germplasm. *Genomics,* Vol. 92, No. 6, (October 2008), pp.478-487, ISSN 0888-7543

Abdurakhmonov, I.Y.; Saha, S.; Jenkins, J.N.; Buriev, Z.T.; Shermatov, S.E.; Scheffler, B.E.; Pepper, A.E.; Yu, J.Z.; Kohel, R.J. & Abdukarimov, A. (2009). Linkage Disequilibrium Based Association Mapping of Fiber Quality Traits in G. hirsutum L. Variety Germplasm. *Genetica,* Vol. 136, No. 3, (July 2009), pp. 401-417, ISSN 1573-6857

Abdurakhmonov, I.Y.; Buriev, Z.T.; Shermatov, Sh.E.; Kushanov, F.N.; Makamov, A.; Shopulatov, U.; Turaev, O.; Norov, T.; Akhmedov, Ch.; Mirzaakhmedov, M. & Abdukarimov, A. (2011). Utilization of Natural Diversity in Upland Cotton (G. hirsutum) *Germplasm Collection for Pyramiding Genes via Marker-Assisted Selection Program,* Proceedings of 5th meeting of Asian Cotton Research and Development Network, *Lahore, Pakistan, February 23-25, 2011*

Adams, K.L.; Percifield, R. & Wendel, J.F. (2004). Organ-specific Silencing of Duplicated Genes in a Newly Synthesized Cotton Allotetraploid. *Genetics,* Vol. 168, No. 4, (December 2004), pp. 2217-2226, ISSN 0016-6731

Alvarez, I. & Wendel, J.F. (2006). Cryptic Interspecific Introgression and Genetic Differentiation Within *Gossypium aridum* (Malvaceae) and Its Relatives. *Evolution,* Vol. 60, No. 3, (March 2006), pp. 505-517, ISSN 1558-5646

Arpat, A.B.; Waugh, M.; Sullivan, J.P.; Gonzales, M.; Frisch, D.; Main, D.; Wood, T.; Leslie, A.; Wing, R.A. & Wilkins, T.A. (2004). Functional Genomics of Cell Elongation in Developing Cotton Fibers. *Plant Molecular Biology,* Vol. 54, No. 6, (April 2004), pp. 911–929, ISSN 1573-5028

Azmat, M.A. & Khan, A.A. (2010). Assessment of Genetic Diversity among the Varieties of Gossypium arboreum and Gossypium hirsutum through Random Amplification of Polymorphic DNA (RAPD) and Simple Sequence Repeat (SSR) Markers. *Pakistan Journal of Botany,* Vol. 42, No. 5, (October 2010), pp. 3173-3181, ISSN 0556-3321

Barroso, P.A.V.; Hoffman, L.V.; de Freitas, R.B.; de Batista, A. C.E.; Alves, M.F.; Silva, U.C. & de Andrade, F.P. (2010). In situ Conservation and Genetic Diversity of Three Populations of *Gossypium mustelinum* Myers ex Watt. *Genetic Resources and Crop Evolution,* Vol. 57, No. 3, (March 2010), pp. 343-349, ISSN 0925-9864

Bertini, C. H. C. M.; Schuster, I.; Sediyama, T.; Barros E.G. & Moreira, M.A. (2006). Characterization and Genetic Diversity Analysis of Cotton Cultivars Using Microsatellites. *Genetics* and Molecular Biology, Vol. 29, No. 2, (2006), pp. 321-329, ISSN 1415-4757

Blenda, A.J.; Scheffler, B.; Scheffler, M.; Palmer, J.M.; Lacape, J.Z.; Yu, C.; Jesudurai, S.; Jung, S.; Muthukumar, P.; Yellambalase, S.; Ficklin, M.; Staton, R.; Eshelman, R.; Ulloa, M.; Saha, S.; Burr, B.; Liu, S.; Zhang, T.; Fang, D.; Pepper, A.; Kumpatla, S.; Jacobs, J.; Tomkins, J.; Cantrell, R. & Main, D. (2006). CMD: A Cotton Microsatellite Database Resource for *Gossypium* Genomics. *BMC Genomics,* Vol. 7, No. 1, (December 2006), pp. 132, ISSN 1471-2164

Boopathi, N.M.; Gopikrishnan, A.; Selvam, N.J.; Ravikesavan, R.; Iyanar, K.; Muthuraman, S. & Saravanan, N. (2008). Genetic Diversity Assessment of G. barbadense Accessions to Widen Cotton (*Gossypium* spp.) Gene Pool for Improved Fibre Quality. *Journal of Cotton Research and Development*, Vol. 22, No. 2, (July 2008), pp. 135-138, ISSN 0972-8619

Brubaker, C.L.; Paterson, A.H. & Wendel, J.F. (1999). Comparative Genetic Mapping of Allotetraploid Cotton and Its Diploid Progenitors. *Genome*, Vol. 42, No. 2, (April 1999), pp. 184-203, ISSN 1480-3321

Campbell, B.T.; Saha, S.; Percy, R.; Frelichowski, J.; Jenkins, J.N.; Park, W.; Mayee, C.D.; Gotmare, V.; Dessauw, D.; Gband, M.; Du, X.; Jia, Y.; Constable, G.; Dillon, S.; Abdurakhmonov, I.Y.; Abdukarimov, A.; Rizaeva, S.M.; Abdullaev, A.A.; Barrose, P.A.V.; Padua, J.G.; Hoffman, L.V. & Podolnaya, L. (2010). Status of Global Cotton Germplasm Resources. *Crop Science*, Vol. 50, No. 4, (July 2010), pp. 1161-1179, ISSN 1435-0653

Chaudhary, L.; Sindhu, A.; Kumar, M. & Saini, M. (2010). Estimation of Genetic Divergence Among Some Cotton Varieties by RAPD Analysis. Journal of Plant Breeding and Crop Science, Vol. 2, No. 3, (March 2010), pp. 039-043, ISSN 2006-9758

Chen, Z.J.; Scheffler, B.E.; Dennis, E.; Triplett, B.A.; Zhang, T.; Guo, W.; Chen, X.; Stelly, D.M.; Rabinowicz, P.D.; Town, C.D.; Arioli, T.; Brubaker, C.; Cantrell, R.G.; Lacape, J.M.; Ulloa, M.; Chee, P.; Gingle, A.R.; Haigler, C.H.; Percy, R.; Saha, S.; Wilkins, T.; Wright, R.J.; Van Deynze, A.; Zhu, Y.; Yu, S.; Abdurakhmonov, I.; Katageri, I.; Kumar, P.A.; Mehboob-Ur-Rahman, Z.Y.; Yu, J.Z.; Kohel, R.J.; Wendel, J.F. & Paterson, A.H. (2007). Toward Sequencing Cotton (*Gossypium*) Genomes. *Plant Physioogyl*, Vol. 145, No. 4, (December 2007), pp. 1303-1310, ISSN 1532-2548

De Almeida, V.C.; Hoffman, L.V.; Yokomizo G.K.I.; da Costa, J.N.; Giband, M. & Barroso, P.A.V. (2009). In situ Genetic Characterization of *Gossypium barbadense* Populations from the States of Para and Amapa, Brazil. *Pesquisa Agropecuaria Brasileira*, Vol. 44, No. 7, (July 2009), pp. 719-725, ISSN 0100-204X

Dejoode, D.R. & Wendel, J.F. (1992). Genetic Diversity and Origin of the Hawaiian-Islands Cotton, *Gossypium tomentosum. American Journal of Botany*, Vol. 79, No. 11, (November 1992), pp. 1311-1319, ISSN 0002-9122

Dighe, N.D.; Robinson, A.F.; Bell, A.A.; Menz, M.A.; Cantrell, R.G & Stelly, D.M. (2009). Linkage Mapping of Resistance to Reniform Nematode in Cotton Following Introgression from *Gossypium longicalyx* (Hutch.& Lee). *Crop Science*, Vol. 49, No. 4, (July-August 2009), pp.1151-1164, ISSN 0011-183X

Endrizzi, J.; Turcotte, E. & Kohel, R. (1985). Genetics, Cytology and Evolution of *Gossypium. Advances in Genetics*, Vol. 23, pp. 271-375, ISBN 9780120176236

Esbroeck, G. & Bowman, D. (1998). Cotton Germplasm Diversity and its Importance to Cultivar Development. *Journal of Cotton Science*, Vol. 2, No. 3, (July 1998), pp. 121-129, ISSN 1523-6919.

Feng, Ch.; Ulloa, M.; Perez-M, C. & Stewart, J.McD. (2011). Distribution and Molecular Diversity of Arborescent *Gossypium* Species. *Botany*, Vol. 89, No. 9, (September 2011), pp.615-624, ISSN 1916-2790

Fryxell, P.A. (1979). Natural History of the Cotton Tribe. Texas A&M University Press, College Station, Texas, USA

Fryxell, P.A. (1992). A revised taxonomic interpretation of *Gossypium* L. (Malvaceae). *Rheedea*, Vol. 2, No. 1, (January 1992), pp. 108-165, ISSN: 0971-2313

Gingle, A.R.; Yang, H; Chee, P.W.; May, O.L.; Rong, J.; Bowman, D.T.; Lubbers, E.L.; Day, J.L.; & Paterson, A.H. (2006). An Integrated Web Resource for Cotton. *Crop Science*, Vol. 46, No. 5, (September 2006), pp. 1998-2007, ISSN 1435-0653

Guo, W.; Sang, Z.Q.; Zhou, B.L. & Zhang, T. (2007). Genetic Relationships of D-genome Species Based on Two Types of EST-SSR Markers Derived from *G. arboreum* and *G. raimondii* in *Gossypium*. *Plant Science*, Vol. 172, No. 4, (April 2007), pp. 808-814, ISSN 0168-9452

Guo, W.Z.; Zhou, B.L.; Yang, L.M.; Wang, W. & Zhang, T.Z. (2006). Genetic Diversity of Landraces in Gossypium arboretum L. Rase *sinense* Assessed with Simple Sequence Repeat Markers. *Journal of Integrative Plant Biology*, Vol. 48, No. 9, (September 2006), pp. 1008-1017, ISSN 1744-7909

Gutierrez, O.A.; Basu, S.; Saha, S.; Jenkins, J.N.; Shoemaker, D.B.; Cheatham, C.L. & McCarty, J.C. (2002). Genetic Distance Among Selected Cotton Genotypes and Its Relationship With F2 Performance. *Crop Science*, Vol. 42, No. 6, (November–December 2002), pp. 1841-1847, ISSN 1435-0653

Hawkins, J.S.; Pleasants, J. & Wendel, J.F. (2005). Identification of AFLP Markers That Discriminate Between Cultivated Cotton and the Hawaiian Island Endemic, *Gossypium tomentosum* Nuttall ex Seeman. *Genetic Resources* and *Crop Evolution*, Vol. 52, No. 8, (December 2005), pp. 1069-1078, ISSN 1573-5109

Hinze, L.L.; Kohel, R.J.; Campbell, B.T. & Percy, R.G. (2011). Variability in Four Diverse Cotton (Gossypium hirsutum L.) Germplasm Populations. *Genetic Resources and Crop Evolution*, Vol. 58, No. 4, (April 2011), pp. 561-570, ISSN 0925-9864

Ibragimov, P.S.; Avtonomov, V.A.; Amanturdiev, A.B.; Namazov, S.E.; Zaurov, D.E.; Molnar, T.J.; Eisenman, S.W.; Orton, T.J.; Funk, C.R.; Percival, J. & Edward, A. (2008). Uzbek Scientific Research Institute of Cotton Breeding and Seed Production: Breeding and Germplasm Resources. *Journal of Cotton Science*, Vol. 12, No. 2, (April 2008), pp. 62-72, ISSN 1524-3303

International Cotton Advisory Committee [ICAC]. (2011). Small Increase in Global cotton consumption expected in 2011/2012. Press Release, (September 2011), Washington, DC, USA.

Iqbal, J.; Reddy, O.U.K.; El-Zik, K.M. & Pepper, A.E. (2001). A Genetic Bottleneck in the 'Evolution Under Domestication' of Upland Cotton *Gossypium hirsutum* L. Examined Using DNA Fingerprinting. *Theoretical and Applied Genetics*, Vol. 103, No. 4, (September 2001), pp. 547-554, ISSN 1432-2242

Iqbal, M.J.; Aziz, N.; Saeed, N.A.; Zafar, Y. & Malik, K.A. (1997). Genetic Diversity Evaluation of Some Elite Cotton Varieties by RAPD Analysis. *Theoretical and Applied Genetics*, Vol. 94, No. 1, (July 1997), pp. 139-144, ISSN 1432-2242

Kalivas, A.; Xanthopoulos, F.; Kehagia, O. & Tsaftaris, A.S. (2011). Agronomic Characterization, Genetic Diversity and Association Analysis of Cotton Cultivars using Simple Sequence Repeat Molecular Markers. *Genetics and Molecular Research*, Vol. 10, No. 1, (February 2011), pp. 208-217, ISSN 1676-5680

Kalivas, A.; Xanthopoulos, F.; Kehagia, O. & Tsaftaris, A.S. (2011). Agronomic Characterization, Genetic Diversity and Association Analysis of Cotton Cuitivars using Simple Sequence Repeat Molecular Markers. *Genetics and Molecular Research*, Vol. 10, No. 1, (February 2011), pp. 208-217, ISSN 1676-5680

Kantartzi, S.K. & Stewart, J.McD. (2008). Association Analysis of Fibre Traits in *Gossypium arboreum* Accessions. *Plant Breeding*, Vol. 127, No. 2, (April 2008), pp. 173-179, ISSN 1439-0523

Kantartzi, S.K.; Ulloa, M.; Sacks, E. & Stewart, J.M. (2009). Assessing Genetic Diversity in Gossypium arboreum L. Cultivars Using Genomic and EST-derived Microsatellites. *Genetica*, Vol. 136, No.1, (May 2009), pp. 141-147, ISSN *1573-6857*

Kebede, H.; Burow, G.; Dani, R.G. & Allen, R.D. (2007). A-genome Cotton as a Source of Genetic Variability for Upland Cotton (*Gossypium hirsutum*). *Genetic Resources and Crop Evolution*, Vol. 54, No. 4, (June 2007), pp. 885-895, ISSN 1573-5109

Khan, A.I.; Fu, Y. & Khan, I. (2009). Genetic Diversity of Pakistani Cotton Cultivars as Revealed by Simple Sequence Repeat Markers. *Communications in Biometry and Crop Science*, Vol. 4, No. 1, (2009), pp. 21-30, ISSN 1896-0782

Kump, K.L.; Bradbury, P.J.; Wisser, R.J.; Buckler, E.S.; Belcher, A.R.; Oropeza-Rosas, M.A.; Zwonitzer, J.C.; Kresovich, S.; McMullen, M.D.; Ware, D.; Balint-Kurti, P.J. & Holland, J.B. (2011). Genome-Wide Association Study of Quantitative Resistance to Southern Leaf Blight in the Maize Nested Association Mapping Population. *Nature Genetics*, Vol. 43, No. 2, (January 2011), pp. 163-168, ISSN 1546-1718

Lacape, J.M.; Dessauw, D.; Rajab, M.; Noyer, J.L. & Hau, B. (2007). Microsatellite Diversity in Tetraploid *Gossypium* germplasm: Assembling a Highly Informative Genotyping Set of Cotton SSRs. *Molecular Breeding*, Vol. 19, No. 1, (January 2007), pp. 45-58, ISSN 1572-9788

Liu, D.; Guo, X.; Lin, Z.; Nie, Y. & Zhang, X. (2006). Genetic Diversity of Asian Cotton (*Gossypium arboreum* L.) in China Evaluated by Microsatellite Analysis. *Genetic Resources and Crop Evolution*, Vol. 53, No. 6, (September 2006), pp. 1145-1152, ISSN 1573-5109

Liu, S.; Cantrell, R.G.; McCarty, J.C.J. & Stewart, J.M. (2000). Simple Sequence Repeat-based Assessment of Genetic Diversity in Cotton Race Stock Accessions. *Crop Science*, Vol. 40, No. 5, (September 2000), pp. 1459-1469, ISSN 1435-0653

Liu, Z.; Zhang, Y.; Zhang, S.; Deng, K.; Dong, S. & Ren, Z. (2011). Genetic Diversity Analysis of Various Red Spider Mite-Resistant Upland Cotton Cultivars Based on RAPD. *African Journal of Biotechnology*, Vol. 10, No. 18, (May 2011), pp. 3515-3520, ISSN 1684-5315

Luconge, E.; Herselman, L. & Labuschagne, M.T. (2007). Genetic Diversity of Tanzanian Cotton (Gossypium hirsutum L.) Revealed by AFLP Analysis, *Proceedings of African Crop Science Conference*, Vol. 8, pp. 773-776, ISSN 1023-070X, El Minia, Egypt, October 27-31, 2007

Mahmood, Z.; Raheel, F.; Dasti, A.A.; Shahzadi, S.; Athar, M. & Qayyum, M. (2009). Genetic Diversity Analysis of the Species of Gossypium by using RAPD Markers. *African Journal of Biotechnology*, Vol. 8, No. 16, (August2009), pp. 3691-3697, ISSN 1684-5315

Mauer, F. M. (1954). Origin and Taxonomy of Cotton, In: *Cotton*, 383, Academy of Sciences of USSR, Tashkent, Uzbekistan (In Russian)

McCarty, J.C.Jr. & Jenkins, J.N. (1993). Registration of 79 Day-neutral Primitive Cotton Germplasm Lines. Crop Science. Vol. 33, No. 2, (March-April 1993), pp. 351, ISSN 1435-0653

McCarty, J.C.Jr.; Jenkins, J.N.; Parrott, W.L. & Greech, R.G. (1979). The Conversion of Photoperiodic Primitive Race Stocks of Cotton to Day-neutral Stocks. *Mississippi Agricultural & Forestry Experiment Station Research Report*, Vol. 4, No. 19, (July 1979), pp. 1-4, ISSN 0147-2186

Multani, D.S. & Lyon, B.R. (1995). Genetic Fingerprinting of Australian Cotton Cultivars With RAPD Markers. *Genome*, Vol. 38, No. 5, (October 2005), pp. 1005-1008, ISSN 1480-3321

Patel, J.P.; Fougat, R.S. & Jadeja, G.C. (2009). Genetic Diversity Analysis in some Elite Desi Cotton Cultivars of Gossypium herbaceum and G. arboreum and Genetic Purity Testing of Their Hybrids through Microsatellite Markers. *International Journal of Plant Sciences*, Vol. 4, No. 2, (May2009), pp. 464-470, ISSN 0973-1547

Paterson, A.; Boman, R.; Brown, S.; Chee, P.; Gannaway, J.; Gingle, A.; May, O. & Smith, C. (2004). Reducing the Genetic Vulnerability of Cotton. *Crop Science*, Vol. 44, No. 6, (November-December 2004), pp. 1900-1901, ISSN 1435-0653.

Percival, A.E.; Stewart, J.M. & Wendel, J.F. (1999). Taxonomy and Germplasm Resources, In: *Cotton: Origin, History, Technology and Production*, C.W. Smith and J.T. Cothren, (Ed), 33-63, ISBN 978-0-471-18045-6 John Wiley, New York.

Phillips, L.L. (1964). Segregation in New Allopolyploids of *Gossypium*. V. Multivalent Formation in New World x Asiatic and New World x wild American Hexaploid. *American Journal of Botany*, Vol. 51, No. 3, (March 1964), pp. 324-329, ISSN 0002-9122

Pillay, M. & Myers, G.O. (1999). Genetic Diversity in Cotton Assessed by Variation in Ribosomal RNA Genes and AFLP Markers. *Crop Science*, Vol. 39, No. 6, (November 1999), pp. 1881-1886, ISSN 1435-0653

Poland, J.A.; Bradbury, P.J.; Buckler, E.S. & Nelson, R.J. (2011). Genome-Wide Nested Association Mapping of Quantitative Resistance to Northern Leaf Blight in Maize. *Proceedings of the National Academy of Sciences of the USA*, Vol. 108, No. 17, (April 2011), pp. 6893-6898, ISSN 0027-8424

Rahman, M.; Yasmin, T.; Tabassam, N.; Ullah, I.; Asif, M. & Zafar, Y. (2008). Studying the Extent of Genetic Diversity among Gossypium arboreum L. Genotypes/Cultivars using DNA fingerprinting. *Genetic Resources and Crop Evolution*, Vol. 55, No. 3, (May2008), pp. 331-339, ISSN 0925-9864

Rana, M.K.; Singh, V.P. & Bhat, K.V. (2005). Assessment of Genetic Diversity in Upland Cotton (*Gossypium hirsutum L.*) Breeding Lines by Using Amplified Fragment Length Polymorphism (AFLP) Markers and Morphological Characteristics. *Genetic Resources and Crop Evolution*, Vol. 52, No. 8, (December 2005), pp. 989-997, ISSN 1573-5109

Reddy, U.K.; Rong, J.; Nimmakayala, P.; Vajja, G.; Rahman, M.A.; Yu, J.; Soliman, Kh.M.; Heller-Uszynska, K.; Kilian, A. & Paterson, A.H. (2011). Use of Diversity Arrays Technology Markers for Integration into a Cotton Reference Map and Anchoring to a Recombinant Inbred Line Map. *Genome*, Vol. 54. No. 5, (May 2011), pp. 349-359, ISSN 0831-2796

Roberts, P.A., Ulloa, M. (2010). Introgression of Root-knot Nematode Resistance into Tetraploid Cotton. *Crop Science*, Vol.50, No.3, (May 2010), pp. 940–951, ISSN 1435-0653

Robinson, A.F.; Bell, A.A.; Dighe, N.D.; Menz, M.A.; Nichols, R.L. & Stelly, D.M. (2007). Introgression of Resistance to Nematode *Rotylenchulus reniformis* into Upland

Cotton (*Gossypium hirsutum*) from *Gossypium longicalyx*. *Crop Science*, Vol. 47, No. 5, (September-October 2007), pp. 1865-1877, ISSN 0011-183X

Ruan, Y.L.; Llewellyn, D.J. and Furbank, R.T. (2003). Suppression of Sucrose Synthase Gene Expression Represses Cotton Fiber Cell Initiation, Elongation, and Seed Development. *The Plant Cell*, Vol. 15, No. 4, (April 2003), pp. 952-964, ISSN 1531-298X

Rungis, D.; Llewellyn, D.; Dennis, E.S. & Lyon, B.R. (2005). Simple Sequence Repeat (SSR) Markers Reveal Low Levels of Polymorphism Between Cotton (*Gossypium hirsutum* L.) Cultivars. *Australian Journal of Agricultural Research*, Vol. 56, No. 3, (March 23, 2005), pp. 301-307, ISSN 1836-5795

Sapkal, D.R.; Sutar, S.R.; Thakre, P.B.; Patil, B.R.; Paterson, A.H. & Waghmare, V.N. (2011). Genetic Diversity Analysis of Maintainer and Restorer Accessions in Upland Cotton (*Gossypium hirsutum* L.). *Journal of Plant Biochemistry and Biotechnology*, Vol.20, No.1, (January-June 2011), pp. 20-28, ISSN 0971-7811

Stelly, D.M.; Lacape, J.M.; Dessauw, D.E.G.A.; Kohel, R.; Mergeai, G.; Saha., S.; Sanamyan, M.; Abdurakhmonov, I.Y.; Zhang, T.; Wang, K.; Zhou, B. & Frelichowski, J. (2007). International Genetic, Cytogenetic and Germplasm Resources for Cotton Genomics and Genetic Improvement, *Proceedings World Cotton Research Conference-4*, P. 2154, ISBN, Lubbock, Texas, USA, September 10-14, 2007

Stich, B. & Melchinger, A.E. (2010). An Introduction to Association Mapping in Plants. *CAB Reviews: Perspectives in Agriculture, Veterinary Science, Nutrition and Natural Resources*, Vol. 5, No. 039, (May 2010), ISSN 1749-8848

Sunilkumar, G.; Campbell, L.M.; Puckhaber, L.; Stipanovic, R.D. & Rathore K.S. (2006). Engineering Cottonseed for Use in Human Nutrition by Tissue-specific Reduction of Toxic Gossypol. *Proceedings of the National Academy of Sciences of the USA*. Vol. 103, No. 48, (November 2006), pp. 18054-18059, ISSN 0027-8424.

Tatineni, V.; Cantrell, R.G. & Davis, D.D. (1996). Genetic Diversity in Elite Cotton Germplasm Determined by Morphological Characteristics and RAPD. *Crop Science*, Vol. 36, No. 1, (January-February 1996), pp. 186-192, ISSN 1435-0653

Tian, F.; Bradbury, P.J.; Brown, P.J.; Hung, H.; Sun, Q.; Flint-Garcia, S.; Rocheford, T.R.; McMullen, M.D.; Holland, J.B. & Buckler, E.S. (2011). Genome-Wide Association Study of Leaf Architecture in the Maize Nested Association Mapping Population. *Nature Genetics*, Vol. 43, No. 2, (January 2011), pp. 159-162, ISSN 1546-1718

Tiwari, R.S. & Stewart, J.McD. (2008). Molecular Diversity and Determination of Hybridization Among the G-genome *Gossypium* Species, In: *Summaries of Arkansas Cotton Research* 2008, Research Series 573, (October 2008), D.M. Oosterhuis, (Ed.), pp. 30-33, ISSN 1942-160X

Ulloa, M.; Stewart, J.M.; Garcia-C, E.A.; Goday, A. S.; Gaytan-M, A. & Acosta, N.S. (2006). Cotton Genetic Resources in the Western States of Mexico: *in situ* Conservation Status and Germplasm Collection for *ex situ* Preservation. *Genetic Resources* and *Crop Evolution*, Vol. 53, No. 4, (June 2006), pp. 653-668, ISSN 1573-5109

Ulloa, M.; Brubaker, C.; & Chee, P. (2007). Cotton. In: *Genome Mapping & Molecular Breeding, Vol. 6: Technical Crops*. C. Kole (ed), pp.1-49. Springer, ISBN 9783540345374, Heidelberg, Berlin, New York, Tokyo

Ulloa, M.; Percy, R.; Zhang, J.; Hutmacher, R. B.; Wright S. D. & Davis., R. M. (2009). Registration of four Pima cotton germplasm lines having good levels of *Fusarium* wilt race 4 resistance with moderate yields, and good fibers. *J. Plant Registr.* Vol. 3, No. 2, (May 2009), pp. 198-202, ISSN 1936-5209

Van Deynze, A.; Stoffel, K.; Lee, M.; Wilkins, T.A; Kozik, A.; Cantrell, R.G.; Yu, J. Z.; Kohel, R. J. & Stelly, D. M. (2009). Sampling nucleotide diversity in cotton. BMC Plant Biol., Vol. 9, (October 2009), pp.125, ISSN 1471-2229

Wallace, T.P.; Bowman, D.; Campbell, B.T.; Chee, P.; Gutierrez, O.A.; Kohel, R.J.; McCarty, J.; Myers, G.; Percy, R.; Robinson, F.; Smith,W.; Stelly, D.M.; Stewart, J.M.; Thaxton, P.; Ulloa, M. & Weaver, D.B. (2009). Status of the USA Cotton and Crop Vulnerability. *Genetic Resources and Crop Evolution*, Vol. 56, No. 4, (June 2009), pp. 507-532, ISSN 0925-9864

Wendel, J.F. (1995). Cotton, In: *Evolution of Crop Plants*, S. Simmonds and J. Smartt, (Ed), 358-366, 1st ed Longman, ISBN 978-0-582-08643-2, London

Wendel, J.F. & Albert, V.A. (1992). Phylogenetics of the Cotton Genus (*Gossypium*) - Character State Weighted Parsimony Analysis of Chloroplast-DNA Restriction Site Data and Its Systematic and Biogeographic Implications. *Systematic Botany*, Vol. 17, No. 1, (January-March 1992), pp. 115-143, ISSN 1548-2324

Wendel, J.F. & Percy, R.G. (1990). Allozyme Diversity and Introgression in the Galapagos-Islands Endemic *Gossypium darwinii* and Its Relationship to Continental Gossypium *barbadense*. Biochemical Systematics and *Ecology*, Vol. 18, No. 7-8, (December 2002), pp. 517-528, ISSN 0305-1978

Wendel, J.F.; Brubaker, C.L. & Percival, A.E. (1992). Genetic Diversity in *Gossypium hirsutum* and the Origin of Upland Cotton. *American Journal of Botany*, Vol. 79, No. 11, (November 1992), pp. 1291-1310, ISSN 0002-9122

Wendel, J.F.; Rowley, R. & Stewart, J.M. (1994). Genetic Diversity in and Phylogenetic-relationships of Brazilian Endemic Cotton, *Gossypium mustelinum* (Malvaceae). *Plant Systematics and Evolution*, Vol. 192, No. 1-2, (March 1994), pp. 49-50, ISSN 1615-6110

Westengen, O.T.; Huaman, Z. & Heum, M. (2005). Genetic Diversity and Geographic Pattern in Early South American Cotton Domestication. *Theoretical and Applied Genetics*, Vol. 110, No. 2, (January 2005), pp. 392-402, ISSN 1432-2242

Wu, D.; Fang, X.; Ma, M.; Chen, J. & Zhu, S. (2010). Genetic Relationship and Diversity of the Germplasms in Gossypium barbadense L. from Four Different Countries Using SSR Markers. *Cotton Science*, Vol. 22, No. 2, (March 2010), pp. 104-109, ISSN 1002-7807

Yu, J.; Pressoir, G.; Briggs, W.H.; Vroh, B.I.; Yamasaki, M.; Doebley, J.F.; McMullen, M.D.; Gaut, B.S.; Nielsen, D.M.; Holland, J.B.; Kresovich. S. & Buckler, E.S. (2006). A Unified Mixed-Model Method for Association Mapping That Accounts for Multiple Levels of Relatedness. *Nature Genetics*, Vol. 38, No. 2, (February 2006), pp. 203-208, ISSN 1546-1718

Zeng L.; Meredith, W.R.Jr.; Gutierrez, O.A. & Boykin, D.L. (2009). Identification of Associations Between SSR Markers and Fiber Traits in an Exotic Germplasm Derived From Multiple Crosses Among *Gossypium* Tetraploid Species. *Theoretical and Applied Genetics*, Vol. 119, No. 1, (June 2009), pp. 93-103, ISSN 1432-2242

Zhang, H.B.; Li, Y.; Wang, B & Chee, P.W. (2008). Recent Advances in Cotton Genomics. *International Journal of Plant Genomics*, Vol. 2008, No. 742304, (January 2008), pp. 1-20, ISSN 1687-5389

Zhang, J.; Lu, Y.; Cantrell, R.G. & Hughs, E. (2005a). Molecular Marker Diversity and Field
    Performance in Commercial Cotton Cultivars Evaluated in the Southwestern USA.
    Crop Science, Vol. 45, No. 4, (July 2005), pp. 1483-1490, ISSN 1435-0653
Zhang, J.F.; Lu, Y.Z. & Yu, S.X. (2005b). Cleaved AFLP (cAFLP), a Modified Amplified
    Fragment Length Polymorphism Analysis for Cotton. *Theoretical and Applied
    Genetics,* Vol. 111, No. 7, (November 2005), pp. 1385-1395, ISSN 1432-2242
Zhang, Y.; Wang, X.F.; Li, Z.K.; Zhang, G.Y. & Ma, Z.Y. (2011a). Assessing Genetic Diversity
    of Cotton Cultivars using Genomic and Newly Developed Expressed Sequence
    Tag-Derived Microsatellite Markers. *Genetics and Molecular Research,* Vol. 10, No. 3,
    (July 2011), pp. 1462-1470, ISSN 1676-5680
Zhang, M.; Zheng, X.; Song, S.; Zeng, Q.; Hou, L.; Li, D.; Zhao, J.; Wei, Y.; Li, X.; Luo, M.;
    Xiao, Y.; Luo, X.; Zhang, J.; Xiang, C. & Pei, Y. (2011b). Spatiotemporal
    Manipulation of Auxin Biosynthesis in Cotton Ovule Epidermal Cells Enhances
    Fiber Yield and Quality. *Nature Biotechnology,* Vol. 29, No. 5, (April 2011), pp. 453-
    458, ISSN 1546-1696

4

Exploring Statistical Tools in Measuring Genetic Diversity for Crop Improvement

C. O. Aremu

Department of Crop Production and Soil Science
Ladoke Akintola University of Technology, Ogbomoso, Oyo state
Landmark University, Omu-Aran
Nigeria

1. Introduction

Increase in global numerical population especially in developing nations has gradually led to food shortage and hence increase in poverty. Addressing and tackling the issue and causes of poverty in the developing nations is one major challenge to breeders (Fu and Somers 2009). The different theories of econometircs have identified the human and material resources traceable to poverty, but fail to identify the crop improvement techniques in addressing world food shortage (Baudoin and Mergeai 2001). Crop improvement techniques therefore remains a major concern to plant breeders (Akbar and Kamran, 2006; Aremu *et al*, 2007a). Several factors affect crop improvement for specific or general environment performance. Such factors include climate, weather, soil, edaphic and biological and more importantly crop genotype (Aremu, et al, 2007b). Crop genotypes are composed of different crop forms including inbred or pure lines hybrids, landraces, wildraces germplasm accessions, cultivars or varieties. These crop genotypes have wide and diverse origin and genetic background known as genetic diversity. Genetic diversity study is a major breakthrough in understanding intraspecie crop performance leading to crop improvement (Aremu, 2005). Knowledge of crop performance in genetic diverse population reveals the differences in the nature of genetic materials used.

Genetic diversity studies therefore, is a step wise process through which existing variations in the nature of individual or group of individual crop genotypes are identified using specific statistical method or combination of methods (Christini *et al.* 2009; Warburton and Crossa 2000; Aremu, 2005; Weir 1996). It is expected that the identified variations would form a pattern of genetic relationship useable in grouping genotypes.

Several researchers including breeders have employed different data source and type from diverse crops in their methods to study genetic diversity. Such data source include morphological and agronomic, pedigree, proximate or biochemical and molecular data (Aremu, *et al.*, 2007a in cowpea; Liu *et al.*, 2000 in cotton; Mostafa *et al.*, 2011 in wheat; Adewale *et al.*, 2010 in African Yam bean; Christine *et al.*, 2009 in bentgrass.

The choice of statistical method to be used is dependent on the achievable objectives laid out in the studies. This chapter reveals the underlying importance of genetic diversity and reviews useable statistical techniques for identifying and grouping genotypes for intraspecies crop improvement.

2. Need for germplasm resource in genetic diversity preservation

Crop genotypes sourced as germplasm accessions, landraces, breeding lines, wild species, have rich and variable genetic integrity explorable for breeding programmes. The first step of any meaningful breeding programme is to identify crop plants that exhibit exploitable variation for the trait(s) of interest. However, these genetic diverse crops are under threat. Continuous hybridization and crossing systems have reduced the genetic variations in cropping programmes and leave a dearth in harvesting and utilization of novel crop types with exploitable traits. Also, the continuous threat or loss of genetic diversity as a result of replacement of landraces, wild species and other primitive term of crop species by exotic high- yielding varieties remains an insurmountable problem to plant breeders. Another major source of loss of genetic diversity is by changes and or increase in population size, resulting in land use acts promoting deforestation, wars, industrialization, urbanization and other factors. According to Brown (1989), preservation of genetic diversity is possible when genetic or germplasm resource is realized as the most precious asset in conserving genetic diversity. Germplasm therefore is an essential resource for successful plant breeding. Certain areas of the world exhibit high level of genetic variability for crops (Vavilov, 1950). Falconer and Mackay (1996); Eivazi et al; (2007); reported that such areas are considered as regions or center of genetic diversity. Therefore genetic diversity in crop may be associated with the origin of the crop. This is supported by Christine et al. (2009), who reported genetic diversity to be associated with origin. Potter and Doyle, (1992) reported Tropical Africa to be the centre of diversity for African yam bean. Van Bueningen and Busch (1997), reported genetic diversity of wheat to be centered in North America. Ariyo and Odulaja (1991), found correlation between genetic diversity and eco-geographic background in okro. Some grouping methods in genetic diversity studies identified origin and geographical diversity not important in measuring genetic diversity. Nair et al. (1998) discovered diversity in sugarcane not to be associated with origin. Aremu et al. (2007a), discovered that center of origin is not a measure of genetic diversity in cowpea. If crop origin is somewhat not important in the measure of genetic diversity a resource centre is therefore needed to preserve and maintain the wide genetic sources exploitable in breeding programmes. Genetic relationship and diversity are useful for developing germplasm conservation strategies and utilization of crop genetic resources. The use of genetic diversity resource centre cannot be under estimated as earlier discussed.

3. Importance of genetic diversity studies

Study on genetic diversity is critical to success in plant breeding. It provides information about the quantum of genetic divergence and serves a platform for specific breeding objectives (Thompson et al, 1998). It identifies parental combinations exploitable to create segregating progenies with maximum genetic potential for further selection, as proven by Akoroda (1987), Weir, (1996), Liu et al.(2000); Dje et al.(2000), (Aremu et al, 2007b). Genetic diversity exposes the genetic variability in diverse populations and provides justification for introgression and ideotype breeding programmes to enhance crop performance. Mostafa al et. (2011), postulated that genetic diversity studies provides the understanding of genetic relationships among populations and hence directs assigning lines to specific heterogeneous groups useable in identification of parents and hence choice selection for hybridization. Choice of parent has been identified to be the first basic step in meaningful breeding programme (Akoroda 1987); (Aremu et al. 2007a); (Islam 2004), (Rahim *et al*, 2010). Furthermore, the choice of parent selection in diversity studies is valuable because it is a means of creating useful variations in subsequent progenies.); Dje *et al.* (2000), discovered that the higher the genetic distance between parents, the higher the heterosis in the developed progenies. Hence the heterotic progenies can be further hybridized and selections based on transgressive segregation. Akbar and Kamran, (2006). exploited this parental selection technique in wheat breeding program through hybridization. Mostafa *et al.* (2011), investigated genetic distance among 36 winter wheat genotypes cultivated in different regions of Iran using principal component analysis and discovered five major groups in the genotypes to distantly related. Comprehensive and significant emphasis are made by researchers especially plant breeders on the analysis of genetic diversity in a number of field crops white and yellow yam, (Akoroda, 1987); cowpea, (Adewale and Aremu, 2010); African yam bean, (Baudoin and Mergeai 2001); Flax, (Mohammadi et al. 2010); wheat, (Mostafa *et al.* 2011) and several other crops.

The diversity studies on these crops at their respective primitive levels (Landrace, wildtype, accessions, lines *etc*) led to the development of their widely distributed cultivars and varieties with proven characteristics based on stability and adaptability of performance with consistent tolerance to adverse weather conditions and resistant to diseases around the world. Fu and Somers (2009) supported that the use of identified wheat parents resistant to environmental stress under different growing conditions has led to increased world wheat production. The early report of Mohammadi and Prasna (2003) revealed that appropriate parent selection for hybridization in maize using a definite diversity study technique, Bohn *et al* (1999), identified six groups of wheat land races in the Western Iran that can be grown in different geographical locations for improved yield. Martin *et al.,* (2008) discovered 42 cultivars of bentgrass in the mancet city and that only diversity studies would identify reliable and definite cultivar(s) with varietal purity and ensure protection of breeder and consumer rights. Understanding the inter and intra specie genetic relationships as provided by diversity studies has proven to increases hybrid vigor and reduce or avoid re-selection within existing germsplasm. It is worthy of note that existing cultivar populations have narrow genetic bases, hence need for creating variability within and among cultivars using genetic diversity methods.

4. Genetic diversity measurement tools

Genetic diverse populations arising from pure lines, accessions, landraces, wild or weed races are analyzed using a number of methods. Such method can be single or in combination of two or more methods. Franco *et al.* (2001) stressed the need for careful considerations to be made when measuring genetic diversity within and between crop populations in research. Such considerations include:

1. Use of multivariate data collected from morphological or agronomic traits. Such data may effectively display discrete, continuous, binomial ordinal etc. variables.
2. Use of multiple data sets arising from morphological, biochemical and DNA-based collections. The use of such multiple data sets in diversity study helps to reveal the adequacy in terms of strength and constraints in the choice of each of the data sets. The use of multiple data pose some puzzles including can analysis and result interpretation

be based on individual or combined data sets? And more worrisome is the puzzle on how to effectively combine the different data sets and still achieve meaningful result. To provide answers to these puzzles, Wrigley *et al.* (1982), studied phylogenetic relationships among triticeae species using individual and combined analysis of data sets consisting of morphological and DNA-based traits and discovered divergent results in the analysed individual and combined data. The discrepancies in the results may be attributed to the discrete nature of DNA-based data and the continuous variable nature of the morphological data. No wonder Hillis 1987; Chippindale and Wein (1994) suggested the assignment of specific numbers to both quantitative and qualitative traits in morphological, biochemical and molecular data set. In view of this, Pedersen and Seberg (1998) advised that both individual and combined data sets can be analyzed in many possible and meaningful ways to draw conclusions on genetic divergence. In 1999 and 2001, Taba *et al.* and Franco et al., respectively utilized the modified Location Model (MLM) which combines all variables into one multinomial variable called "W" to classify maize accessions from the genetic resource centres of Latin America. Better still, this MLM can combine molecular and morphological data to classify data better than when individual data set is employed. Individual data from morphological, biochemical or molecular data set can be analyzed using one or a combination of techniques. These techniques shall be discussed.

3.   Expected objective to be achieved. This dictates choice of statistical tool in measuring genetic distance and the level of clustering of the intragenic factors in use. Such objective(s) include to determine the quantum of variation and grouping such genotype based on genetic distance, identify action following parental selection. In essence, breeding focus determines applicable method in explaining the nature of genetic divergence.

Variations are recorded in the measurement of genetic diversity in genotype relationships based on genetic distances and grouping populations from individual genotypes such as accessions, lines, wild races etc. The recorded variations are primarily because of the differences in the nature of genetic materials. Therefore, the basis or genetic variance theories which identifies genotype relationships based on genetic distance estimating genetic diversity depends largely on statistical genetic variance theories which identifies genotype relationships based on genetic distance / variance.

5. The use of morphological data to measure genetic distance

Nei, (1973), first defined Genetic distance as the difference between two entities that can be described by allelic variation. This definition was later in 1987, modified to "extent of gene differences among populations that are measured using numerical values. Betterstill, in 1998, Beaumont *et al.*, provided a more comprehensive definition of genetic distance as any quantitative measure of genetic difference at either sequence or allele frequency level calculated between genotype individuals or populations.

The first early work of Anderson (1957), proposed the use of metrogliph and index-score to study the pattern of morphological variations in individual data set. In the early seventies (Singh and Chaudhary 1985) used this method to study morphological variation in green gram. This method uses a range of variations arising from trait such that extent of trait variation is determined by the length of rays on the glyph. The performance of a genotype is adjudged by the value of the index score of that genotype. The score value determine the length of ray which may be small, medium or long Akoroda (1987); Ariyo and Odulaja (1991) and Van Bueningen and Busch (1997), extensively explored the use of metroglyph and index-score to morphological variations in yellow yam, Okro and wild rye accessions respectively.

Similar to metroglyph and the score index is Euclidian Distance (ED) measurement. According to Nei (1987), Euclidian distance measures similarity between two genotypes, populations or individuals using using statistical measures where two individuals i and j, having observations on morphological traits (p) denoted by $x_1, x_2, x_3, \ldots\ldots x_n$ and $y_1, y_2, \ldots\ldots y_n$ for i and j individuals respectively.

Metroglyph and index-score methods measures genetic distance by use of morphological traits. Euclidian distance measurements utilize both morphological and molecular based marker data sets. Smith et al. (1991), applied the following statistic to measure ED.

$$d_{ij} = \varepsilon[(T_{1(i)} - T_{2(i)}^2)/\sigma^2 T_{(i)}]^{1/2}.$$

Where T_1 and T_2 are the values of the ith trait for 1 lines and 2 and $\sigma^2 T(i)$ is the variance for the ith trait over all the lines used. Much later, Weir (1996) developed a formula for calculating genetic distance to be.

$$d(I,j) = [(x_1 - y_1)^2 + (x_2 - y_2)^2 + \ldots..(x_p - y_p)^2]^{1/2}$$

where i and j is the ED between two individuals lines having morphological traits (p)

$x_1, x_2 \ldots\ldots x_p$ is the traits for i individuals and

$y_1, y_2 \ldots\ldots x_p$ is the traits for j individuals

from here, the individual character distances are summed and then divided by the total number of characters scored in both individuals. ED measurement allows the use of both qualitative and quantitative data several workers identified genotype distances using ED. Van Bueningen and Busch (1997) in wheat, smith et al, 1987 in sorghum and Ajmone – Marsan (1998) in maize.

6. The use of molecular data in measuring genetic distances

The advent and explorations in molecular genetics led to a better definition of Euclidean distance by Beaumont et al; (1998) to mean a quantitative measure of genetic difference calculated between individuals, populations or species at DNA sequence level or allele frequency level.

Various genetic distance measurements are proposed for analyzing DNA-based data for the purpose of genetic diversity studies. Powel et al. (1996), identified different DNA-based marker techniques to include Random Amplified Polymorphic DNA (RAPD), Amplified Fragment Length Polymorphism (AFLP), Restriction Fragment Length Polymorphic (RFLPs) and the most recent Simple Sequence Repeats (SSR) and Microsatellite (MT) of single nucleotide polymorphism (SNPs). The above nucleotide differences can be used effectively to run individual or combined data sets of morphological, biochemical or DNA based data. For DNA based data, where the amplification products are equated to alleles, the allele frequencies can be calculated and the genetic distance between i and j individuals estimated as follows.

$$d(ij) = 1 < \left[\sum_{}^{n} (X_{ai} - X_{aj}) \right]^{1/r}$$

Where X_{ai} is frequency of the allele a for individual I, and n is the number of alleles per loci; r is the constant based on the coefficient used. In its simple form, i.e $r = 1$, genetic distance can be calculated as:

$$d\int (ij) = 1/2 \left[\sum_{}^{n} (X_{ai} - X_{aj}) \right]$$

Where $r = 2$, $d(i,j)$ is referred to as Rogers (1972) measure of distance (RD), where

$$RD_{ij} = 1/2[\Sigma(x_{ai}-x_{aj})^2]^{1/2}$$

Where allele frequencies are to be calculated for some of the molecular markers, the data must first generate a binary matrix for statistical analysis. Binary data has been long and widely used before the advent of molecular marker data to measure genetic distance by Rogers (1972); Nei and Chesser (1983) coefficient and known as GD_{MR} and GD_{NL} respectively.

In the use of any given statistical formula to determine genetic diversity in molecular based data, one specific problem usually encountered is the failure of some genotypes to show amplification for some primer pairs. Robinson and Harris (1999) noted that lack of amplification may be due to "null alleles". Most often, it is difficult to ascribe lack of amplification to "null allele". It is therefore the reposed confidence of the researcher, that a "null allele" status of a genotype will not be considered as missing data during computation of genetic similarity- distance matrix so as to avoid gross error during result interpretation.

DNA based marker data have been successfully used to measure genetic distance in some crops (Pritchard et al. (2000) in pigeon pea; Beaumont et al. (1998) in wheat; Franco et al., (2001) in maize; Dje et al. (2000) in Sorghum.

7. Grouping techniques in measuring genetic diversity

Genetic relationship among and with breeding materials can be identified and classified using multivariate grouping methods. The use of established multivariate statistical algorithms is important in classifying breeding materials from germplasm, accessions, lines, and other races into distinct and variable groups depending on genotype performance. Such groups can be resistant to diseases, earliness in maturity, reduced canopy drought resistant etc. The widely used techniques irrespective of the data source (morphological, biochemical and molecular marker data) are cluster analysis, Principal Component Analysis (PCA), Principal Coordinate Analysis (PCOA) Canonical Correlation and Multidimensional Scaling (MDS).

Cluster analysis presents patterns of relationships between genotypes and hierarchical mutually exclusive grouping such that similar descriptions are mathematically gathered into same cluster (Hair *et al.* 1995); (Aremu 2005). Cluster analysis have five methods namely unweighted paired group method using centroids (UPGMA and UPGMC), Single Linkages (SLCA), Complete Linkage (CLCA) and Median Linkage (MLCA). UPGMA and UPAMC provide more accurate grouping information on breeding materials used in accordance with pedigrees and calculated results found most consistent with known heterotic groups than the other clusters (Aremu *et al.,* (2007a).

Principal components, canonical and multidimentional analyses are used to derive a 2-or 3-dimensonal scatter plot of individuals such that the geometrical distances among individual genotypes reflect the genetic distances among them. Wiley (1981), defined principal component as a reduced data form which clarify the relationship between breeding materials into interpretable fewer dimensions to form new variables. These new variables are visualized as different non correlating groups.

Principal components analysis first determines Eigen values which explain the amount of total variation displayed on the component axes. It is expected that the first 3 axes will explain a large sum of the variations captured by the genotypes. Cluster and principal component analysis can be jointly used to explain the variations in breeding materials in genetic diversity studies.

8. Conclusion

Genetic diversity studies is in no measure the first basic step in meaningful breeding programme and therefore require accurate and reliable means for estimation. Data sets sourced can morphological biochemical several workers successfully utilized various statistical tools in analysis diverse data sets and identified two major framework to really explain divergence in genotype performance. Genetic distance among and within individual data sets can be conveniently determined using specific tools while classificatory and cluster analysis require principal component and polymorphic sequence tools. Since each data set provide different molecular type of information, based marker data set is visualized to provide more reliable differentiate information on the genotypes. Analysis of data sets can be complex. Many software packages are available. There is still a need for a comprehensive and user-friendly software packages that would integrate different data set for analysis and generate reliable and useable information about genetic relationship. Equally important in genetic diversity studies is the need for a genetic resource centre. Studies should incorporate utilization of genetic diversity information in developing genetic resource centre accessible to breeders.

9. Acknowledgement

Many thanks to Ibirinde Olalekan for the secretariat assistance. I also appreciate Olayinka Olabode the Head of Department of Agronomy LAUTECH for the technical contributions given to this chapter.

10. References

Adewale, B.D.,Kehinde, O.B., Aremu, C. O. Popoola, and Dumet, J. 2010. Seed metrics for genetic and shape determinations in African yam bean [Fabaceae] (*Sphenostylis stenocarpa)* African Journal of Plant Science Vol. 4(4): 107-115

Ajmone-Marsan P., Carstiglioni, P., Fusari, F., Kuiper, M. and Motto, M. 1998. Genetic diversity and its relationship to hybrid performance in maize as revealed by RFLP and AFLP markers. Theor. Appl. Genet. 96:219-227.

Akbar, A.A. and Kamran, M. (2006). Relationship among yield components and selection criteria for yield improvement of Safflower – Carthamustinctorious L. J. Appl. Sci. 6: 2853-2855.

Akoroda, M.O. 1987. Principal component analysised metroglyph of variation among Nigerian yellow yam. Euphytica 32:565-573.

Aremu C.O. 2005. Diversity selection and genotypes Environment interaction in cowpea unpublished Ph.D Thesis. University of Agriculture, Abeokuta, Nigeria. P. 210.

Aremu, C.O., Adebayo M.A., and Adeniji O.T. 2008: Seasonal performance of cowpea in humid tropics using GGE biplot analysis world jour. of Biol.. Research. 1: (1) 8-13.

Aremu, C.O., Adebayo, M.A., Ariyo O.J., and Adewale B.D. 2007b. Classification of genetic diversity and choice of parents for hydridization in cowpea vigna unguiculata (L) walip for humid savanna ecology. African journal of biotechnology 6: (20) 2333-2339.

Aremu, C.O., Adebayo, M.A., Oyegunle, M. and Ariyo, J.O.2007a. TRhe relative discriminatory abilities measuring Genotype by environment interaction in soybean (Glycine max). Agricultural journ. 2 (2).: 210-215

Ariyo, O.J. and Odulaja, A. 1991. Numerical analysis of variation among accessions of Okra. Ann. Bot. 67:527-531.

Barrett, B.A. and Kidwell, K.K. 1998. AFLP-based genetic diversity assessment among wheat cultivars from the pacific Northwest crop Sci. 38:1261-1271.

Baudoin, J. and Mergeai G. 2001. Yam bean sphenostylis stendcarpa. In: R.H. Raemaekers (ed) crop production in tropical Africa Directorate general for international (DGIC) Brussels, Belgium. Pp. 372 – 377.

Bearmont, M.A., Ibrahim, K.M., Boursot, P. and Bruqord, M.W. 1998. Measuring genetic distance. P. 315-325. In A. karp et al. (ed) molecular tools for screening biodiversity. Chapman and Hall, London.

Bohn, M.,H.F. Hutz and A.E.Melchinegr. 1999. Genetic similarities among winter wheat cultivars determined on the basis of RFLPs, AFLPs ans SSRs and their used ion predicting progeny variance. Crop Sci. 39; 228-237.

Brown, A.H.D. 1989. The case for core collection. P. 135-156. In A.H.D England Brown et al. (ed). The use of plant genetic resources. Cambridge Univ. press Cambridge.

Brown-Guedira, G.L; Thompson J. Ajnelson, R.L. and Warburton M.L: 2000. Evaluation of genetic diversity of soybean introduction and North American ancestors using RAPD and SSR markers crop Sci. 40:815-823.

Bueninger, L.T. and Busch, Ritt. 1997. Genetic diversity among North American spring wheat cultivars. Analysis of the coefficient of parentage matrix. Crop Sc. 37:570-579.

Chippindale, P.T. and Weins J.I. 1994. Weighting portioning and combining characters in phylogenetic analyses. Syst. Biol. 43:273-287.

Christine, Joshua, H., William, A., Stacy, A. 2009. Genetic diversity of creeping bentgrass cultivars using SSR markers. Intern. Turfgrass Soc. Research Jour. 11.

Dje, Y, Hevretz, M., Letebure, C. and Vekemans, X. 2000. Assessment of genetic diversity within and among germplasm accessions in cultivated sorghum using microsatellite marked theor. Appl. Gent. 100:918-925.

Eivazi, A.R, Naghavi, M.R, Hajheidari, M. Mohammadi, S.A, Majidi, S.A, Salakdeh, I. and Mardi, M. 2007. assessing wheat genetic diversity using quality traits, amplified fragment length polymorphism simple sequence repeats and proteome analysis Ann. Appl. Biol. 152:81-91.

Falconer, D.S. and Mackay T.F. 1996. Introduction to Quantitative Genetics,Longman, Harlow.

Franco, J., Crossa, J., Ribaot, M., Betran, J., Warburton, M. land Khairallah, M. 2001. A method for combinary molecular markers and phenotypic attributes for classifying plant genotype Theor-Appl. Gent. 103: 944-952.

Fu, Y. and Somers D. 2009. Genomo-wide reduction of genetic diversity in wheat breeding Crop Sci. 49:161-168.

Hair, J.R., Anderson, R.E., Tatham, R.L., and Black, W.C. 1995. Multivariatev data analysis with Readings. 4th edition, Prentice- Hall, Englewood Cliffs, NJ.

Hillis, D.M. 1987. Molecular versus morphological approaches to systematics. Annu. Rev. Ecol. Systems. 18:23-42.

Islam. M.R. 2004. Genetic diversity in irrigated rice pak. Jorn of Biol. Sci. 2:226-226.

Joshi, B.K, Mudwari, A., Bhatta, M.R, Ferrara, G.O. 2004. Genetic diversity in Nepalese wheat cultivars based on agro-morphological traits and coefficient of parentage. Nep Agric Res. J. 5:7-17.

Liu, S., Cantrell, R.G., Mccarty, J.C. and Stewart, M.D. 2000. simple sequence repeat based assessment of genetic diversity in cotton race accessions. Crop Sci. 40:1459-1469.

Martin, E., Cravero, V., Esposito A, Lozez F, Milanebi, L. and Cointry, E. 2008. Identification of markers linked to agronomic traits in globe artichoke. Aust. J. Crop Sci. 1:43-46.

Mohammadi A.A, Saeidi, G. and Arzuni, G. 2010. Genetic analysis of some agronomic traits in flax. Aust. J. Crop sci. 4:343-352.

Mostafa K., Mohammad, H. and Mohammad, M. 2011 genetic diversity of wheat genotype baspdon cluster and principal component analyses for breeding strategies Australian Jour. of Crop Sc. 5 (1): 17-24.

Nei, M. 1973. Analysis of gene diversity in subdivided populations. Proc. Natil. Acad. Sci. (USA) 70. 3321-3323.

Nei, M. and Chesser, R.K. (1983). Estimation of fixation indices and gene diversities. Ann. Hum. Genet. 47: 253.-259

Pedersen, G. and Seberg, O. 1998. Moleculesvs Morphology. P. 359-365. In A. karp et al (ed) Molecular tools for screening Biodiversity.. Chapman and Hall, London.

Potter, D. and Doyle, J. J. 1992. Origin of African yam bean (Sphenostylisstenocarpa, Leguminosae): evidence from morphology, isozymes,chloroplast DNA and Linguistics. Eco. Bot. 46: 276-292.

Powell, W., Morgante, M., Andre, C., Hanafey, M., Vogel, J.Tinjey, S. and Rafalsky, A. 1996. The comparison of RFLP, RAPD, AFLP and SSR markers for germplasm analysis. Mol.breed. 2: 225-238.

Rahim, M.A., Mia, A.A, Mahmud, F., Zeba, N. and Afrin, K. 2010. Genetic variability, character association and genetic divergence in murgbean platn Omic 3:1-6.

Rogers, J.S. 1972. Measures of genetic similarity and genetic distance studies in genetics. VII. Univ. Tex. Publ. 2713: 145-153.

Singh R.K. and Chaudhary B.D. 1985. Biometrical methods in quantitative genetic analysis. Kalyani publishers, New Delhi. India. P 38.

Smith, J.S., Paszkiewics, S. Smith,O.S.and Schaoffer, J. 1987. Electorphoretic, chromatographic and genetic techniques diversity among corn hybrids. P. 187-20. In Proc. Am.Seed Trade Assoc.Washington DC.

Smith, J.S., Smith, O.S., Boven, S.L., Tenburg, R.A.and Wall, S.J. 1991. The description and assessment of distances between inbred lines of maize III: A revised scheme for the testing of distinctiveness between inbred lines utilizing DNA RFLPs. Maydica, 36. 213-226.

Swoftord D.L, Olsen, G.J. Wadell, P.J. and Hillis D.M. 1996. Phylogenetic inference P. 407-514. In D.M. Hillis et al (ed) Molecualr systematics. 2nd edition Sinaver Associates, Sunderland, M.A.

Taba, S., Diaz, J., Franco, J., Crossa, J.and Eberhart, S.A. 1999. A core subject of LAMP, from the Latin American maize project. CD-ROM, CIMMYT, Mexico, D.F., Mexico.

Thompson, J.A., Nelson, R.L. and Vodkin, L.O.. 1998. Identification of diverse soybean germplasm using RAPD markers. Crop Sci. 38: 1348-1355.

Van Bueningen, L.T. and Busch, R.H. 1997. Genetic diversity among North American spring wheat cultivars: I. Analysis of the coefficient of parentage matrix. Crop Sci. 37: 570-579.

Vavilov, N.I. 1950. The origin, variation, immunity and breeding of cultivated plants. Chronica botanica, B. chronica Botanica company, Waltham, Massachusetts. P 364.

Warburton, M. and Crossa, J. 2000. Data analysis in the CIMMYT. Applied Biotechnology Center for fingerprinting and Genetic Diversity Studies. CIMMYT, Mexico.

Weir, B.S. 1996. Intraspecific differentiation P. 385-403. in D.M. Hillis et al. (ed). Molecular systematics 2nd edition sunderlands M.A.

Wiley, E.O., 1981. Phylogenetics: The theory and practice of phylogenetics and systemic.John Wiley, New York.

Wrigley, C.W, Autran, J.C. and Bushuk, W. 1982. Identification of cereal varieties by gel electorphoresis of the grain proteins. Adv. Cereal. Sci. Techolog. 5:211-259.

5

Genetic Structure and Diversity of Brazilian Tree Species from Forest Fragments and Riparian Woods

Danielle Cristina Gregorio da Silva et al.*
Universidade Estadual do Norte do Paraná
Brazil

1. Introduction

Historical patterns of human occupation in Brazilian Neotropical Region, featured by deforestation for urbanization, economic exploitation and agriculture, have changed the Atlantic Forest landscape to a collection of fragments. Nowadays this biome is characterized for being highly fragmented but still possessing one of the highest rates of species diversity in the world.

Understanding the genetic structure of populations that occur in forest remnants is fundamentally necessary to establish efficient strategies for the re-composition, management, and conservation programs. For such, it is necessary not only to understand the genetic diversity of a species, but also, how this diversity is distributed within and between forest populations. Notably, a considerable part, if not the majority, of Brazilian Atlantic Forest fragments are linked to rivers or streams, once the policy applied in Brazil regarding conservation in agricultural areas favours the maintenance of legal reserves in proximity of water sources. The vegetation of river margins are subjected to flooding, a strong limiting factor which can lead to local adaptation. These ecological and landscape characteristics may have important outcomes to the genetic diversity of tree populations.

In this chapter, we assembled information from review and research papers of impact in this study area intending to raise knowledge to assist conservation initiatives of Brazilian Atlantic Forest fragments and riparian woods. We plan to broach fundamental concepts regarding the historical fragmentation process in Brazilian Neotropical Region, the effects of fragmentation upon the genetic diversity of forest remnants, and the local adaptation to seasonally variable river levels. This discussion is not intended to be a summary of the existing literature in the theme, but to address important information concerning genetic diversity of neotropical tree species, focusing in the results of eleven years of research on species frequently used in reforestation of legal reserves in Southern Brazil.

* Mayra Costa da Cruz Gallo de Carvalho[1], Cristiano Medri[2], Moacyr Eurípedes Medri[3], Claudete de Fátima Ruas[3], Eduardo Augusto Ruas[3], Paulo Maurício Ruas[3]
[1]*Empresa Brasileira de Pesquisa Agropecuária, Brazil*
[2]*Universidade Estadual do Norte do Paraná, Brazil*
[3]*Universidade Estadual de Londrina, Brazil*

2. The Neotropical floristic kingdom

The word *Neotropic* (from the Greek *neos* = "new") refers to the tropical region of the American continent (Antonelli & Sanmartín, 2011), or "New World" – a term coined by Peter Martyr d' Anghiera in 1493, shortly after Christopher Columbus's first voyage to the Americas (O'Gorman, 1972). As currently defined (Schultz, 2005), the Neotropical kingdom extends from central Mexico, in the north, to southern Brazil, in the south, including Central America, the Caribbean islands and most of South America.

In the Neotropics, equatorial and tropical climates predominate with low climatic seasonality, when compared with kingdoms with cold and temperate climates. Precipitation and annual mean temperatures are generally high, but there is great regional variation (Antonelli & Sanmartín, 2011). Before human colonization, the rain forest of Amazonia accounted for about one third of the entire South American continent. There are, however, several other terrestrial biomes in the Neotropics that are noteworthy for their size and ecological importance, such as the Cerrado and the Atlantic forest of eastern Brazil (Antonelli & Sanmartín, 2011).

The outstanding species richness found today in the Neotropics has remained elusive in our understanding of the evolution of life on Earth (Antonelli & Sanmartín, 2011). Comprising around 90,000–110,000 species of seed plants, the Neotropics alone harbours about 37% of the world's species, more than tropical Africa (30,000–35,000 spp.) and tropical Asia and Oceania combined (40,000–82,000 spp.) (Govaerts, 2001; Thomas, 1999). Sanmartín and Antonelli punctuated the factors that can explain this high species richness (Antonelli & Sanmartín 2011 and references therein). They are related 1) with the geographical position of Central and South America, resulting largely in tropical and equatorial climates; 2) edaphic heterogeneity; 3) biotic interactions that promote speciation mechanisms; 4) relatively stable environments over time, resulting in very ancient ecosystems that conserve niches; 5) adaptive radiation favoured by the great ability of dispersal of flowering plants; 6) geographic isolation (only 3.0 million years ago, South America became connected to North and Central America by the Isthmus of Panama); 7) climatic fluctuations of Pleistocene, leading to formation of refuges for many isolated populations and allopatric speciation; 8) uplift of the Andean cordillera, occurred largely in the last 25 million years; and 9) the profound change in the vast hydrological systems, especially in the Amazon region.

During millions of years there was a synergism between gradual and slow climate changes and speciation, giving time for natural selection and other evolutionary tolls to play their role. The final result of this long process of genesis and evolution of the Neotropical region, is that most of the Neotropical countries, such as Brazil, Colombia, Ecuador and Peru, are on the higher positions of any ranking of species richness. Brazil, in particular, occupies the first position in such rankings and is therefore considered the most Megadiverse country.

Neotropical kingdom is divided into Caribbean, Guayana Highlands, Amazonian, Brazilian and Andean region (Takhtajan, 1986). Brazilian region is, in turn, subdivided into the provinces of the Caatinga, the highlands of central Brazil, Chaco, Atlantic, and Paraná (Takhtajan, 1986). In the Brazilian territory, major biomes are spread over these provinces, such as the Caatinga, Cerrado, Atlantic Forest, Pantanal and Pampas. Even in the case of a predominantly tropical region, where, typically, there is less thermal seasonality and spatial climate variations than in temperate regions, there are important environmental variations that create significant intra-regional heterogeneity. This fact can be explained by: 1) large latitudinal variation found, the northern boundary of the region is located around 3 ° S, and the south boundary, around 33 ° S latitude; 2) great altitudinal variation, with elevations ranging from sea level to mountainous regions that can reach approximately 2900 meters of altitude; 3) wide edaphic variation, the result of different soil genesis processes over time and space and 4) continentality, which determines a lower overall level of humidity and higher thermic amplitude, the most distant of the Atlantic Ocean. This great heterogeneity, combined with the inherent environmental characteristics of tropical regions that favor the development of life, and added to the climate and geological history of the entire Neotropic, which favored isolation and speciation, have provided to these biomes, in terms of floristic and physiognomy, a high species richness, high endemism and great structural complexity.

According to the Brazilian Institute of Geography and Statistics - IBGE (2004), Atlantic Forest biome constitutes the extra-Amazonian forest large set of South America, formed by rainforests (dense, open and mixed) and seasonal (deciduous and semideciduous). It comprises an environmental complex that includes mountain ranges, plateaus, valleys and plains of the entire eastern Brazilian Atlantic continental band. In southern and southeastern Brazil, it expands westward, reaching the borders of Paraguay and Argentina, also advancing on the southern highlands of Brazil, reaching the state of Rio Grande do Sul.

3. The history of fragmentation of the Atlantic forest biome

The Atlantic forest is the most uncharacterized Brazilian biome (IBGE, 2004). Since the beginning of European colonization, from 1500, several economic cycles of exploration occurred, generating successive impacts in a growing area (Dean, 1996). Brazil has the most diverse flora of the planet and also had its name inspired in a tree, Brazil-wood (*Caesalpinia echinata* Lam), a typical species of coastal forests of the southeast. The first cycle of the ancient Portuguese colony was the exploitation of this species, highly valued for producing a resin which conferred a reddish color to fabrics. Later, with the largest reserves of this wood already exhausted, other exploitation or agricultural cycles began, such as the sugar cane, gold, and ultimately, coffee. They all contributed strongly to clearance and degradation of new areas (Dean, 1996). From the late 19th century and throughout the 20th century, new development and national integration projects have come and settled a consistent process of industrialization and urbanization exactly in the area originally occupied by the Atlantic Forest. Nowadays, these urban areas have the highest population densities and lead the economic activities in the country (IBGE, 2004). An estimated 112 million people live in this area, which accounts for 61% of the population of Brazil (IBGE, 2007). The current results are the almost complete loss of primary forests and a continuous process of destruction of existing remnants, which place the Atlantic Forest biome in unworthy position in the world: as one of the most endangered ecosystems (SOS Mata Atlantica and Instituto Nacional de Pesquisas Espaciais (INPE), 2011). In sum, predatory economic cycles and projects of development and national integration have led, throughout the biome area, to the expansion of the agricultural frontier, the establishment of industrial activities, mining and power generation and intense and disorganized urbanization process, causing the destruction of, approximately, 92% of the original vegetation, which had, in 1500, 1,315,460 square kilometers (Fundação SOS Mata Atlântica and INPE, 2011). The remaining 8% are highly fragmented, separated by a matrix that includes pastures, crops, water reservoirs, industrial plants, mining and urban areas, especially on the margins of water bodies and mountain areas with highly tilted ground. Even so, one should not underestimate the importance of these fragments. For example, in terms of species richness, occur, even today, in the Atlantic Forest biome, about 20.000 species of vascular plants, of which 6.000 are endemic (Fundação SOS Mata Atlântica and INPE, 2011).

If on one hand the historical processes of economic development led to a predatory pattern of destruction with the formation of relictual fragments of sizes and levels of isolation and different anthropic impact, on the other, the Brazilian environmental legislation, considered one of the most advanced of the planet, through his last Forest Code, established in 1965 and improved in recent decades, defined areas of permanent preservation and legal reserve (Medeiros et al., 2004). The first has the function of conservation of ecosystem services, encompassing riparian forests, river headwaters, hills tops and mountains, hillsides with slopes greater than 45 °, "restingas" and mangroves. The second has the goal of biodiversity conservation and must have, at least, 20% of the property area in the southern and southeastern Brazil, reaching 80% of the area in the Amazon region. Thus, the historical process that led to the intense fragmentation of the Atlantic Forest biome has suffered, in recent decades, the influence of an environmental legislation more effective for conservation. The result is that most of the remaining fragments can be found in permanent preservation areas and legal reserves, especially in riparian areas and steep slopes with unstable soils.

At this time, Brazilian society is burned with a heated debate, in the Brazilian National Congress, on the proposed changes to the Forest Code. In one side, the sectors of society and policymakers linked to the economically strong Brazilian agricultural sector require the flexibility of code, aiming to reduce the need for recovery and conservation of areas of permanent preservation and legal reserve, in order to result in the release of more land for agriculture. In the other side, the Brazilian Society for Science Progress, urban sectors of society and policymakers linked to the environmental movement strongly oppose the proposed changes. There is a concern that, if these changes are approved, the fragmentation process of natural areas and the destruction of relictual fragments will be intensified, which would result in ecological and evolutionary consequences for populations of many species present in these natural areas.

4. Genetic diversity of tree species

4.1 Fundamental concepts

The genetic variability contained in plant species may occur in distinct levels: 1) species within ecosystems; 2) populations within species; and 3) individuals within populations of a species. The genetic structure of a species can be defined as the distribution of the genetic variability within and between its populations as a direct result of the combination among mutation, migration, selection and genetic drift. Also, many tree species harbor effective mechanisms that allow the dispersion of alleles, enabling them to maintain high levels of genetic variability within their populations (Hamrick et al., 1979; Hamrick, 1983; Loveless and Hamrick, 1987). Studying several natural tropical tree populations, Hamrick (1983), concluded that the genetic variability within natural populations is directly linked to their mating system, pollen dispersion syndrome, seed propagation and also by their effective population size. These factors are also related to the geographic distribution and the type of community in which such species naturally occur.

As stated above, understanding the genetic structure of populations that occur in forest remnants is fundamentally necessary to establish adequate criteria in to which these populations will play a role in the re-composition of degraded ecosystems (Kageyama, 1987). For such, it is necessary not only to understand the genetic diversity of a species, but also, how this variability is distributed within and between their populations present in the disturbed areas. To such intention many statistical tools have since been developed to measure, qualify and partition this genetic variability.

In 1951, Sewal Wright established one of the main components of the distribution of genetic diversity in natural populations: the partition of the endogamy coefficient within and among populations. In his method Wright was able to determine how the endogamy coefficient determine not only the level of crossing between close related individuals within a population but also how it can be related to the differentiation of multiple populations and the overall adaptability of a metapopulation. Also he was able to demonstrate how this partition could be directly linked to the matting system present within each species. His method partitioned the components of the endogamy coefficient f into three distinct coefficients: F_{IS}, which is mainly used to measure the degree of crossing between closely related individuals within a population; F_{ST}, which can be considered an estimative of the endogamy level among populations. Although such concept might be strange, Wright was able to determine that this endogamy between populations was equivalent to determining the genetic relatedness among these populations; and F_{IT}, which represent the endogamy level present in the whole metapopulation and correspond to the overall adaptability measured for the sum of populations. We can also consider that F_{IS} represents the endogamy level related to the reproductive system present within a species, F_{ST} to be the endogamy level due to the partition of the population into subpopulations, and F_{IT} the endogamy level related to the reproductive system and the subdivision of all subpopulations.

This concept of partitioning the f statistic developed by Wright is highly important when we consider the genetic study of the fragmented population as it gives an overview of how the genetic variability is distributed within and among the subpopulations of a species in a determined area and permits us not only to infer the level of fragmentation within a metapopulation but also main type of reproduction present in a species.

When considering the genetic pattern of natural populations, we need also to know specific patterns of genetic richness within each population to answer a wide range of questions like: degree of conservation, percentage of variation within each specific population, differences in genetic diversity and degree of heterozygosity. This being said, we have to consider other important measures of population genetics.

The percentage of polymorphic loci is one of the simplest measurements of the genetic variation that can be used to evaluate the genetic variability present in a population. As the name says, it shows the number of polymorphic loci present in a population in relation to all the amplified loci obtained with the DNA-based Markers.

Another form to evaluate the genetic variability present within a population is to obtain information about its genetic diversity. Initially geneticist borrowed the method of calculating such diversity from our friends Ecologists. The Shannon and Weaver (1949) index (H) was employed in ecology to measure the diversity of species within a given area and later was adapted to measure the genetic diversity within the studied populations. Later on, a more specific index was developed to measure this genetic diversity eliminating some of the bias that was generated when adapting the Shannon index for the evaluation of the genetic data. The Nei´s genetic diversity (1973) was developed as a specific way to measure the population genetic diversity using data obtained specifically with the DNA-based markers. This index was able to measure more accurately the degree of genetic variation within each population and presented different considerations when analyzing the data obtained by dominant (there is no way to differentiate the recessive alleles) codominant (all alleles are differentiated, and each pair of primers is considered to amplify only one molecular locus) markers. For dominant (H) data Nei´s genetic diversity is analyzed in terms of within population gene diversity (Hs) and the total gene diversity present in the pool of populations studied (Ht). But as said above, for codominant marker more detailed information can be given by this statistical index, dividing this statistic into observed (Ho) and expected (He) heterozygosities, making it possible to calculate the excess or deficit of heterozygotes within each population, gene flow and inferences of genetic bottlenecks and genetic drift.

Considering that in the traditional method of calculation the genetic variability was based on the assumption that the populations were in Hardy-Weinberg equilibrium, some of the inferences obtained for this population presented a significant bias. As a good example we can consider the inference obtained with the Nei´s statistic for genetic diversity, the observed heterozygosity (Ho) was compared to the expected heterozygosity (He) which was obtained as the pattern of distribution of all the alleles amplified if the population was in HW equilibrium. To eliminate this bias the Baeysian statistic method was adapted to the analysis of molecular data, promoting a revolution in the parameters that can be calculated using data obtained by the DNA-based markers. The Bayesian method is characterized by the use of the posterior probability to infer the likelihood of occurrence of a particular event. In this method all the assumptions, like HW equilibrium, are discarded and the obtained results with this method are compared with a chain (algorithms - Markov Chain of Monte Carlo MCMC and Metropolis Hastings) of results that come from the analyses of the same data. This method of chained analyses repeats itself in tandem until the obtained "chain" stabilizes and yields a result that come close to the real pattern of genetic variation present in natural populations. With this method some parameters like population bottlenecks, attribution of genotypes to specific populations, pattern of gene flow and migration, and attribution of individual paternity, allowing inferences on the population genetics based on DNA-based markers more complete and trustworthy.

4.2 Consequences of habitat fragmentation to the genetic diversity of tree species

Habitat fragmentation is one of the most important and well diffused consequences to the anthropic soil use dynamics (Brooks et al. 2002). It is characterized by the rupture of landscape unity that initially presented continuity, generating smaller parcels with different dynamics from the original habitat. Such parcels become disconnected from the original biological processes that occurred throughout the area (Dias et al., 2000), behaving like isolated "islands of biodiversity" surrounded by non forest areas (Debinsk and Holt, 2000).

For trees, degradation of primary habitat results from two main processes, fragmentation of forest into patches following clearance, and disturbance of habitat following extraction processes, such as selective logging. Tropical trees are thought to be particularly vulnerable to the effects of habitat degradation due to their demographic and reproductive characteristics (Lowe et al. 2005). Estimations made more than 10 years ago predicted that within fifty years, approximately, twenty-five percent of the vascular plant species would be extinct (Kala, 2000). This loss is still an ongoing process, which is not only linked to the loss of the number of individual plants of a species, but also to the loss of condition of the habitat in which they initially inhabited, as some species cannot persist in small fragments due to alteration in microclimatic conditions and to the intensification of the border effect in small fragments (Lovejoy, 1983).

Tropical trees are predominantly outcrossed, present extensive genetic flow and keep high levels of genetic variability. They frequently experience low density as a consequence of habitat fragmentation, are highly dependent upon animal pollination and present mixed mating systems, so they generally exhibit more genetic diversity among populations than temperate species (Dick et al. 2008).

For neotropical forest species, a reduction in habitat patch size or population density is usually equivalent to a reduction in population size (Lowe et al. 2005), or a genetic bottleneck. The genetic bottleneck leads to a very well known phenomenon in the evolutionary context of habitat disturbance, the genetic drift. Genetic drift can be defined as the sum of random changes in the frequencies of alleles within a population (Futuyma, 2005). This process has as principal outcome the decrease in genetic variability, which can be detected as a drop in the proportion of polymorphic loci and in the number of alleles per locus.

The reduced effective size of populations is also accompanied by the increase in endogamy levels. This can be a result of both the decline in pollen vectors and consequent raise of selfing in species with mixed mating systems, and the increased probability of crossing among relatives, given their close distribution and reduced potential mates. This can lead to a declined heterozigosity, augmented homozygosity and consequent fixation of alleles, independently of their effects over fitness.

The consequences of genetic drift and endogamy are maximized by the isolation of the remaining fragments, resulting in an increased genetic diversity among them. The loss of genetic variability can affect population viability and limit evolutionary opportunities to the populations: they are expected to suffer increased disease and pest susceptibility, loss of incompatibility alleles, fixation of deleterious alleles and decline in fitness (Young et al, 1996).

Lowe and coworkers, based in a computer simulation (Lowe et al. 2005), found that with even relatively low levels of gene flow between remnant populations, loss of diversity can be significantly mitigated by increasing effective population size. Changes in genetic diversity and differentiation following a decrease in population size take a number of generations to become apparent, which is not the case for inbreeding coefficient that increases immediately in the first generation following the occurrence of selfing. Long-lived, historically outcrossing species, such as neotropical trees, are expected to harbour a high genetic load, as deleterious recessive alleles will be masked at multiple heterozygous loci.

Also, although deleterious mutations are expected to be purged by selection over time in these species, mildly deleterious alleles can persist despite strong selection.

Some morpho-physiological and life-history traits could confer differences in plants' vulnerability to the effects of fragmentation. For example, species with long generation times will suffer weaker negative effects of fragmentation than the ones with short generation times. The same for species able to reproduce asexually, that will have an extended time between generations.

Many naturally outcrossed tree species frequently present selfing as a clear result of fragmentation, since low density/fragmented populations tend to be more autogamous than high density populations. In the hypothetical scenario of anthropogenic fragmentation ceasing and landscapes remaining as they are today, the effects on genetic diversity of plants will still be much stronger in the future if mating patterns continue shifting towards selfing (Dick et al. 2008, Aguillar et al. 2008).

Aguillar et al. (2008) showed that outcrossing species, such as neotropical trees, suffer greater losses of alleles and polymorphic loci than non-outcrossing species. For self-incompatible species in particular, this may result in the loss of low frequency self-incompatibility alleles (S) leading to mate limitation and further reduction of effective population size.

Animal pollinated outcrossed species are also strongly negatively affected in terms of effective pollination service and seed production by habitat fragmentation. Considering that more than 98% of tropical trees species are animal pollinated (Bawa 1990), these species are exceptionally vulnerable to fragmentation as a consequence of both, ecological and genetic mechanisms. Not only pollination but also seed dispersion in tropical forests is much more dependent in animals than in wind: more than 70% of all tropical tree species are animal dispersed (Howe and Smallwood 1982). These species share multiple agents and are generally understory (Dick et al. 2008).

For the majority of tropical trees, pollen flow transposes enormously the distances of the seeds flow. Pollen dispersion distances, although dependent on small animals in most cases, can be surprisingly as high as 500m and still higher, a few kilometers, in low density/fragmented populations (reviewed in Dick et al. 2008). Even if sufficient pollen reaches an isolated tree to fertilize all potential ovules, a reduction in diversity of the pollen cloud, due to fewer pollen donors, can reduce population fitness by allowing fertilizations from self, related, or maladapted parents. A reduction in seed set predicted from fragmentation may be due to one or a combination of a lack of pollination (e.g. from loss of pollinators), lack of compatible pollination (from increased self-pollination through restricted pollinator movement or a loss of incompatibility alleles owing to reduced population size), or inbreeding (Lowe et al. 2005).

Removal of primary habitat also usually decreases the probability that migrant seeds will find suitable sites for establishment. Under such circumstances, pioneers or invasive weeds will be favoured and will increase in occurrence. If a species' life history profile is characterized by frequent extinction and colonization events, the metapopulation is under threat of extinction if the two forces are not balanced. Even if site colonization does occur, founder bottlenecks can drastically reduce diversity (Lowe et al. 2005 and citations therein).

Species rarity can also determine its susceptibility to genetic erosion. Because common species have comparatively higher levels of genetic variability than naturally rare species, they are expected to lose more diversity due to recent fragmentation processes (Aguillar et al. 2008). Given the ubiquitous nature of anthropogenic habitat fragmentation in today's landscapes, this is important and of interest to conservation biology as they situate common species in potential risk of genetic erosion, which is counterintuitive to current conservation principles that almost exclusively emphasize efforts on rare or threatened species.

As discussed above, habitat fragmentation has the potential to erode genetic diversity of a species, and the magnitude of its effects is related to the state of several life-history traits. Among the several factors, deserve special attention the compatibility system, mating system, pollination vector, seed dispersal vector, vegetative growth capability, rarity, time elapsed in fragmentation conditions, (reviewed in Aguilar et al. 2008), and sucessional stage, as we a going to discuss below.

5. Local adaptation in tree species

An adaptation is a characteristic that enhances the survival or reproduction of organisms that bear it, relative to alternative character states, especially ancestral condition. The only way to an adaptation to evolve is by means of natural selection, so it can be also stated that adaptation is a characteristic that evolved by natural selection (Futuyma, 2005).

Local adaptation is the set of patterns and processes observed across local populations of the same species connected, at least potentially, by dispersal and gene flow. It is generally the case when resident genotypes in a deme have on average a higher relative fitness in their local habitat than genotypes originated from other habitats (Kawecki and Ebert, 2004 and references therein).

Local adaptation can be observed in a continuous population, in which sampling units are arbitrary, but is more commonly observed in fragmented populations, since they are discrete units of perennial populations in well delimited habitat patches (Kawecki and Ebert, 2004). Considering that the most existing neotropical tree populations have been fragmented by land use and urban occupation, local adaptation is a relevant area of study for neotropical tree species. Even though, most studies on local adaptation in plants are only available for herbaceous plants in temperate regions (Leimu and Fischer, 2008).

We aim to gather information on local adaptation in plants that can be used to evaluate its strength and outcomes for neotropical tree populations, considering their present fragmented character. Our understanding of this topic is mainly based on the papers of Kawecki and Ebert (2004), Savolainen and colleagues (2007), and Leimu and Fischer (2008).

The ability to adapt may be compromised in small populations because of reduced genetic diversity, caused by genetic bottlenecks or founder effects, which have as consequences an increase in genetic drift and inbreeding. In addition to reduced genetic variation and genetic drift, local adaptation can also be constrained by variation in natural selection. Temporal environmental variability may involve opposing selection pressures and thus constrain adaptation. In contrast, spatial heterogeneity of the habitats of plant origin favours selection for reduced dispersal and increases habitat fidelity, which may in turn favour the evolution of local adaptation (reviewed in Leimu and Fischer, 2008).

Reproductive traits have a role in determining the extent of local adaptation in natural populations. Gene flow, for example, can hinder local adaptation. This is true because protected polymorphism in a heterogeneous environment may be maintained even if dispersal results in complete mixing of the gene pool. In such a case demes will not differentiate genetically, i.e. there will be no local adaptation. Nevertheless, the existence of a pattern of local adaptation despite gene flow certifies to the strength of natural selection imposed by particular environmental factors (Kawecki and Ebert, 2004). Also, if local adaptation is constrained by lack of genetic variation, dispersal and gene flow between populations can enhance local adaptation by increasing genetic variation within populations and potential to respond to selection (Leimu and Fischer, 2008).

Spatial environmental heterogeneity favours reduced dispersal and habitat fidelity, which make conditions for local adaptation more favourable. It should be noted that environmental heterogeneity favours the evolution of adaptive phenotypic plasticity. In the absence of costs of and constraints on plasticity, a genotype that in each habitat produces the locally optimal phenotype would become fixed in all demes. Adaptive phenotypic plasticity would thus lead to adaptive phenotypic differentiation, but without underlying genetic differentiation. The failure of the metapopulation to evolve such ideal plasticity is thus a pre-requisite for local adaptation (Kawecki and Ebert, 2004 and reference therein).

Shortlived and self-compatible species tend to be more strongly differentiated at a smaller scale than long-lived and outcrossing species, and so the former are expected to show stronger adaptation to local conditions. Therefore, neotropical tree species, which are long-lived, outcrossing and as a group includes several examples of self incompatibility, are expected to have weak local adaptation (Leimu and Fisher 2008).

Leimu and Fischer (2008) conducted the first quantitative review on local adaptation in plants, assembling papers that reported comparisons of the performance of plants from local and foreign populations. Among these studies, local genotypes performed on average better than foreign genotypes at their site of origin. However, divergent selection favoured locally adapted plants only in less than half of the pair-wise site comparisons. This suggested to them that local adaptation is less widespread than commonly believed. In this study, they also found that local adaptation appeared to be independent of some plant life-history traits, the degree of spatial and temporal habitat heterogeneity, and of the geographic distance between study populations, but was strongly affected by population size. This clear role of population size for the evolution of local adaptation raises considerable doubt on the ability of small plant populations to cope with changing environments. Thus, in the context of fragmentation process in the Neotropic small fragments, featured by low genetic variability as a consequence of genetic drift and endogamy, might not be able to respond to different selective pressures of changing environments and develop local adaptation.

6. Brazilian Atlantic forest fragments: Case studies of species from Tibagi River Basin

6.1 The Tibagi project

In this section of the chapter we intend to describe our experience in population genetic studies of neotropical trees remnants in Brazil. All species that we have studied have a high importance value index or IVI (the sum of relative dominance, relative frequency and relative density) in the Tibagi River basin, Paraná, Brazil and, due to the rapid degradation of ecosystems associated with this river basin, its present occurrence was limited to highly impacted forest remnants. The Tibagi River Basin has a great importance in the economic and social development of one of the Brazilian states of greatest economic impact in the country, however, the degree of landscape devastation that has taken place since the beginning of the last century has been threatening the biodiversity of the ecosystem and even physical and chemical characteristics of its rivers. For these reasons, a group composed of around 51 researchers from the University of Londrina and other research centers, in partnership with COPAT (Consortium for Environmental Protection of the Tibagi River basin) and funded by Klabin S/A and Araucaria Foundation for the Support of Scientific and Technological Development of Paraná, developed the project "Aspects of Fauna and Flora of the Tibagi River basin" or Tibagi Project. The Tibagi Project produced a wealth of valuable information aimed at the recovery and conservation of the river basin as a whole. We limit ourselves here to present the knowledge gained about the influence of two factors on the genetic diversity of populations of neotropical tree species remnants: habitat fragmentation and local adaptation to seasonally flooded river banks.

The Tibagi River basin is composed of 65 direct tributaries and hundreds of sub-tributaries in an area of approximately 25,000 Km^2, covering 54 counties in the Paraná State, Brazil. Its landscape, belonging to the Atlantic Forest biome, presents important climate and soil variations in the north-south axis that allows it to be divided into Upper, Middle, and Lower Tibagi. On the Upper Tibagi, the predominant vegetation is steppe grassy-woody also known as general fields, with patches of Araucaria forest. In the Middle Tibagi, there is a transition zone between rain forest and mixed semideciduous forest with some patches of fields and scrubs. In the Lower Tibagi, before fragmentation, the dominant vegetation was the semi-deciduous forest, which formed a continuum with the dense rain forest of the Brazilian coast. With massive deforestation due to lumber extraction, urbanization and expansion of the agricultural front, only small forest fragments remain in this area adding up to only 2.4% of the original forest cover in the lower Tibagi and 12.7% in the Tibagi River basin as a whole (Ribeiro, 2009).

The riparian forests are plant formations that surround bodies of water and for that reason, are deeply influenced by them. On the other hand, water quality and other physical characteristics of rivers are also heavily influenced by the presence and degree of conservation of the riparian forests. In river banks, where the vegetation is scarce or not present at all, events such as erosions can be up to 30 times more significant than in areas where the riparian vegetation is present. Moreover, it is estimated that 80-90% of sediments and pollutants generated in agricultural areas are filtered by riparian vegetation before reaching the bodies of water (Naiman & Decamps, 1997). From the ecological point of view, these forests work as ecological corridors, linking different ecosystems and thereby promoting gene flow, both by land, and across the river. These are environments in which the diversity is determined by the occurrence of flood events with consequent flooding of the river margins. These environments can be very heterogeneous when flooding events are short, sporadic and with little intensity; however, they can be slightly heterogeneous when these flooding events are constant or very prevalent and intense.

Flooding is the main limiting factor that acts on the riparian forest, since it changes dramatically the physical and chemical characteristics of the soil, which is the main substrate for plant growth. In flooded soil, the diffusion of gases is severely reduced, with a sharp drop in oxygen levels, reduction of gas exchanges, accumulation of toxins, changes in pH and nutrient availability (Drew 1992). When flooding extends to total or partial submergence of the plants, the negative effects on vegetation are even greater due to reduced availability of light and carbon dioxide (Crawford, 1993). Once flood period is over, the soil becomes compacted and a series of injuries from re-aeration process arises due to the accumulation of reactive oxygen species in plant tissues during the flooding period (Crawford, 1993).

However, most of the riparian plant species survive the flooding stress through the development of morphologic, anatomic and physiological changes (collectively referred to as the low oxygen escape syndrome - LOES; Bailey-Serres & Voesenek, 2008) that re-established, at least in part, minimum energy levels for plant survival. Other plant species survive with avoidance strategies, completing their life cycles between periods of flooding. There are also those species that have better relative growth rates in flooded soil. The ability to tolerate periods of flooding gives plants a competitive advantage over those that do not tolerate this stress, which is evidenced by the example of the tolerant species *Cecropia pachystachya*, *Sebastiana commersoniana* and species of the genus *Inga*, which are most often found in areas affected by periodic flooding than in other plant formations, where inter-specific competition limits its distribution. In general, local adaptation in challenging environments may either favour the reduction of genetic variation through natural selection or lead to the expression of phenotypic plasticity of identical genotypes (Stöcklin et al., 2009).

In this light, it seems clear that the recovery strategies and/or conservation of genetic diversity of tree species of the riparian forests on the Tibagi River basin should consider: (i) knowledge about the history of fragmentation and its influence on the genetic structure of the affected populations, (ii) knowledge about the diversity of mechanisms of tolerance to flooding and their possible genetic determination, (iii) and the influence of local adaptation to flooding events on the distribution of genetic variation within and among populations. In order to answer these questions, many common species from Tibagi river basin were studied, regarding flood tolerance and occurrence of morphological, anatomical and physiological changes in response to it. For eight of these species, we used DNA-based molecular markers (RAPD, AFLP and SSR) to obtain estimates of population genetic parameters such as polymorphic loci, total genetic diversity and genetic differentiation within and among populations. The eight sampled species have high importance value index (IVI) in Tibagi River basin (five are among the ten species with highest IVI by region) and cross-pollination system, but vary in regard to gender distribution and dispersal mechanisms for pollen and seeds (Table 1).

In Tibagi River basin, the analysis of the pluviometrical events from 1932 to 1990 showed that there were ten significant flood events in the region in this period, while the fragmentation of the vegetation was more pronounced between 1920 and 1960 (Medri et al., 2002). Considering that the first study published by the group comprised a minimum of 76 years between the collection of plant material and the first pronounced flooding event registered in the region and 83 years in relation to the beginning of the fragmentation process; in the meantime, the considered species advanced at least three generations (Table 1). It was expected at first that: (i) the genetic diversity of populations of degraded areas might have been reduced compared to preserved areas, (ii) the pioneer species have been less affected by the effects of fragmentation than the species of late successional stages, (iii) there has been significant genetic differentiation between fragments and (iv) local adaptation to flooding has caused genetic structure, probably by reducing the diversity in populations that regularly experience stress.

Tree Species	Ss[1]	Ra	MS (SD)[2]	Pd	Sd	Long	Mm	Np	Pp(%)	Ht	Fst	Fis	Ref.
A. sellowiana	Pi/Si	2-4	Ou & Vp (Di)	Bi	Bi	25	SSR	5	77.78-100	0.32-0.48	0.197	0.03(-) 0.33	Oliveira, 2010
							RAPD	2	66.67-67.48	0.17-0.18	0.059	-	Medri et al., 2011
							RAPD	9	35.32-52.38	0.09-0.14	0.495	-	Medri et al., 2011[4]
A. polyneuron	St	50	Ou (He)	Bi	Ab	1200	AFLP	2	88.5-99	0.31-0.37	0.265	-	Damasceno et al., 2011
C. xanthocarpa	St	6-8	Ou (He)	Bi	Bi	60	AFLP	2	92.27-92.82	0.32-0.33	0.180	-	Ruas, 2009
C.gonocarpum	St/Cl	-	Ou (He)	Bi	Bi	-	AFLP	2	79.5-92.1	0.28-0.31	0.300	-	Ruas, 2009
L. divaricata	Pi	2	Ou (He)	Bi	Ab	60	RAPD	2	80.69-86.9	0.9	0.104	-	De Carvalho et al., 2008
M. aquifolium	St/Cl	2	Ou (Un)	-	Bi	60	RAPD	3	72.80-84.51	0.17-0.21	0.218	-	Sahyn et al., 2010
P. rigida	Pi	10	Ou (Mo-Si)	Bi	Ab	100	RAPD	2	91.95	0.94	0.063	-	Silva et al., 2010
							AFLP	8	48.44-82.81	0.17-0.30	0.228	-	Souza, 2011
S. terebinthifolius	Pi	2-4	Ou (Di)	Bi	Bi	10-15	RAPD	2	64.64-70.72	0.168-0.20	0.137	-	Ruas et al., 2011
							SSR	6	87.5-100	0.48-0.62	0.199	0.04(-)0.03	Ruas et al., 2011[4]

[1] Pi: Pioneer; Si: Secundária inicial; St: Secundária tardia; Cl: Clímax.
[2] Ou: Outcrossing; Vp: Vegetatively propagated; Mo: Monoecious; Mo-Si: Monoecious, Self incompatible; Di: Dioecious; He: Hermaphrodite; Un: unknown.
[3] Bi: Biotic; Ab: Abiotic.
[4] In press.

Table 1. Biological features and genetic parameters of the tree species studied in the Tibagi Project initiative. Information on biological traits was obtained in the literature and by personal communications of experts. When more than two populations of a given species were studied, the Fst values are represented as an average of those populations. Ss: Sucessional stages; Ra: Reprodutive age, in years; MS (SD): Mating system (Sex distribution); Pd: Polen dispersal; Sd: Seed dispersal; Long: Longevity, in years; Mm: Molecular marker; Np: Number of populations; Pp: Percentage of polymorphic loci; Ht: heterozigosity or genic diversity, when either codominant or dominant markers are respectively applied; Ref: references.

6.2 Local adaptation of tree species to flooding

In the Amazonian floodplain forests there are some regions where the water column reaches up to 7m with submersion of trees for long periods, and regions where this stress is less intense. Comparing the different areas of flooding, it is possible to observe a significant variation in the occupation pattern of these areas ranging from monospecific to over 150 species ha[-1] (Wittmann et al., 2007 and references therein). It is clear that the stress intensity and duration of flooding periods determine the diversity of species occupying these areas. In the Tibagi River basin, the most critical flooding period is the three wettest summer months (December-January-February), with areas that may remain flooded for a few days to several months depending on the intensity of the phenomenon. Torezan & Silveira (2002), conducted several phytosociological studies in forest fragments along the Tibagi River Basin comparing areas under different intensities of occasional floodings. It was observed that the higher intensity, duration and size of the flooded area, the lower species richness was found. In a fragment with 100% of flooded area, there were 42 species/ha, while in fragments located in non-flooded areas, or with an insignificant area subject to flooding, the number of tree species was often greater than 100/ha, reaching up to 127 in one single area. Thus, we consider as a starting point the hypothesis that the stress of flooding, typically observed in the region, should determine the occupancy of the periodically flooded banks, favoring the establishment of species tolerant to this stress.

Several studies with species that are found in the Tibagi River basin were conducted in a greenhouse to answer this hypothesis. Plant species not found in the wetlands and the species most often found in the wetlands have been challenged in artificial flooding treatments with different intensities and durations. Briefly, the results showed that some tree species naturally distributed in this river basin but that do not occur in flooded areas, did not tolerate the stress of artificial flooding; also, there was a great variation in the response of the species from wetlands. In addition to this information, the provenance trials, conducted to compare the performance of populations of the same species from areas periodically flooded and non flooded, showed that for the studied species, *Luehea divaricata* and *Parapiptadenia rigida*, the plants that originated from populations naturally flooded were more efficient in responding to waterlooging stress (De Carvalho et al., 2008) or tolerated higher levels of stress, which was not tolerated by the plants of the other provenance (Silva et al., 2010). In a similar study with the tree species *Aegiphila sellowiana*, Medri et al. (2011) used plants regenerated from seeds collected in four regions that presented occasional flooding, showing a variation in the response of individual tolerance to waterlogging. In this experiment, 46.7% of the plants died, while the remaining individuals developed morpho-anatomical alterations common to tolerant species, supporting up to 80 days flooding (Medri et al., 2011).

When these results were obtained (2000 - 2002) we believed that there should be a genetic background related to flood tolerance, but the variation in responses observed between the tolerant species pointed to the existence of a large number of genes and/or a strong effect of genotype-environment interaction in determining the tolerance. With these in mind, we used Random Amplified Polymorphic DNA (RAPD) to see if there was a reduction in genetic variation in populations tolerant to flooding, and whether it was possible to detect variation between populations that respond differently to the stress. In our experiments of performance comparison, the percentage of polymorphism observed *in situ* among adults of *Luehea divaricata* and *Parapiptadenia rigida* was greater (approximately 6%) in the areas subjected to periodic flooding than in the highest regions, which is never affected by stress. The studied populations of *L. divaricata* and *P. rigida*, showed genetic differentiation of 10.48 % for the first species and of 6.00% for *P. rigida* (while about 90% of the observed variation was attributed to the variation found within populations). These results suggested that perhaps the expected reduction in genetic diversity of riparian communities has been masked by the balance between different evolutionary forces that could be occurring *in situ*. Considering the proximity of the sampled areas and the fact that both are crossbreeding species, there certainly is gene flow between the flooded and non flooded populations, which in turn must ensure the re-establishment of genetic diversity in each generation. Possibly, the lower inter-specific competition and greater light availability on the river banks may represent, for the pioneer species studied, a more important factor than natural selection caused by flooding, enabling the maintenance of higher levels of diversity in wetlands.

We needed a model that would allow us to access the isolated effect of flooding on the genetic diversity of populations of plants tolerant and intolerant to stress. In this experimental model we use the tree species *A. sellowiana*, which displays a gradation of responses to flooding, ranging from death to survival periods of up to 80 days of stress. When comparing plants of *A. sellowiana* which survived the stress of artificial flooding (tolerant), with plants which died (intolerant), the genetic differentiation around 6% was detected between the two groups and it was not possible to detect important variation in the percentage of polymorphism between both of them (Medri et al., 2011). Given these results, it became clear that the vast phenotypic variation between tolerant and intolerant plants of *A. sellowiana* can be, at least in part, related to the genetic patterns observed when using the DNA-based marker RAPD. A similar situation was found by Bekessy et al. (2003) who used RAPD markers to study genetic variation in populations of the South American tree *Araucaria araucana*. Even though we have yet to acquire the knowledge of which genes determine the stress tolerance of flooding on tree species it is possible that this character is directly influenced by many genes as suggested by Sairam et al. (2008) in their review entitled "Physiology and biochemistry of waterlogging tolerance in plants".

Interestingly, considering the results obtained by the group, especially with the local adaptation experiment, carried out *ex situ* with *A. sellowiana*, it becomes clear that the genetic diversity within and between populations cannot be considered the only factor in the election of sources for the recovery of degraded riparian forests. In short, the RAPD markers used in these studies allowed us to detect consistent results for the three species studied since more than 94% of genetic variation detected can be attributed to variation found within populations, following the pattern usually observed for pioneer species, with preferably cross-fertilization. In the absence of knowledge on the mechanisms of inter-specific diversity of tropical trees on response to flooding, we may conclude, erroneously, based only on estimates of distribution of the genetic diversity, that both the populations that occupy the margins of the flooded rivers and those never flooded could be used as seed sources for the recovery of degraded riparian areas. However, the difference in performance between plants, as measured by morphological parameters, indicate that populations locally adapted to the stress would provide a better material for the recovery of degraded riparian areas.

6.3 Effects of fragmentation in natural populations from Tibagi River Basin

In several years of studies observing the effects of fragmentation and local adaptation on the genetic diversity of the remaining tree populations of the Tibagi River Basin, eight species were evaluated mostly with dominant molecular markers (Table 1). The information gathered here reflect observations for post-fragmented populations, and the number of generations elapsed since the beginning of the fragmentation process is equal to one or two for centenary species, and no more than ten generations for the species with longevity from 20 to 25 years.

As seen among tropical species (Matallana et al., 2005), all species present higher frequencies of cross-fertilization and animals play an important role in pollination and/or seed dispersal. These characteristics can be readily related to the high levels of genetic diversity (or heterozygosity) observed in the sampled populations (Table 1). In addition, pollen and seed dispersal by animals, often observed in tropical species, should provide high gene flow between nearby populations in a continuous gradient, while favouring the genetic differentiation over long distances (Givinish, 1999 and references there in). In such cases, the observed genetic distance between populations was positively correlated with their geographic distances. This hypothesis could be verified from the combined analysis of two studies conducted with *P. rigida*, where two nearby populations (Silva et al., 2010) and eight populations distributed over a long geographic scale (Souza et al., 2011) were evaluated. For nearby populations, the Fst value obtained was only 6.2%, whilst the average F_{ST} between the eight populations was 22.8%, and the correlation between genetic and geographic distance was positive and significant. Similarly, the observed values of F_{ST} between geographically close and distant populations of *A. sellowiana* ranged from 9.56% to 50.26%, respectively (Medri et al. 2011; Oliveira, 2010). For this species in particular, the genetic differentiation between distant populations may be even more pronounced than the one seen for *P. rigida* due to the occurrence of vegetative propagation.

When the genetic diversity among populations is compared between the studied outcrossing species, one can see that there are significant variations among the observed values of F_{ST} (6% - 30%, Table 1), which is suggested to be related to successional stage that each species occupies. Fst values for the initial or secondary pioneer species vary between 6% - 19% and the values for the late secondary or climax species vary between 21% - 30%. In both cases, genetic differentiation among populations assume moderate to high values, which is, in principle, expected to occur among tropical species. Tropical species are often pollinated by animals or have their seeds dispersed by them, often have mixed breeding system in which selfing rates can be changed depending on the environment and are represented by low-density populations (Dick et al. , 2008). These characteristics, together, make the tropical species more dependent on the quality of the ecosystem where they are inserted then the temperate species, and therefore more susceptible to the effects of fragmentation. Under the effect of fragmentation, tropical species suffer a reduction in the availability of pollinators and/or seed dispersers with the reduction in the number of individuals, experiencing a higher proportion of inbreed crosses and genetic drift, with a resulting increase in genetic differentiation between populations.

It is suggested that the observed tendency of lower values of Fst between the pioneer species than among the secondary is a reflection of the life history of these species. The pioneer species represent the first successional stage of the forest, being able to invade areas not yet occupied, including harsh environments and forest edges. Thus, given the ecological role they play, the pioneers are heliophyte, experience rapid growth, lower inter-specific competition, increased investment in reproduction (r selection), a higher number of generations per unit time and aggregate distribution. Fragmentation, therefore, does not cause too negative effects on the pioneer species. In fact, while fragmentation reduces the number of individuals of the species of later successional stages and completely alters the environment they occupy, the pioneer species can be instead favored by this process. Also, the pioneer species from the fragments of the Tibagi River Basin have advanced several generations (10 -15 generations) since the beginning of the fragmentation process. So, after several cycles of cross-fertilization, gene flow between populations may have indeed come relatively high to overcome the effects of fragmentation, allowing lower levels of genetic differentiation among populations.

Unlike the pioneers, the late secondary and climax species tend to have less aggregated spatial distribution, to be ombrophilous and to occupy more specialized niches. These species spend their energy more in the inter-specific competition than in reproduction (k selection) and have slow growth and far fewer generations per unit time. The species of late successional stages, considered here, represent the first or second generation post-fragmentation and thus are believed to reflect the immediate consequences of the fragmentation process. Interestingly, one of the secondary species studied, *Aspidosperma polyneuron*, was found in a continuous distribution, in a plateau followed by a high declivity (Damasceno et al., 2011), and the other, *Maythenus aquifolium*, was found in fragments separated by up to 30 km (Sayhun et al., 2010); however, the F_{ST} values observed in the two situations were similarly high, suggesting that part of the genetic differences found between the populations of these species also linked to adaptive characteristics (Sayun et al. 2010; Damasceno et al., 2011).

Another important information that we could extract from the obtained results for the sampled populations in these regions of lower, middle and upper Tibagi (Ruas et al. 2011 In press; Ruas et al. 2011; Medri et al. 2011 In press; Oliveira et al, 2011) is that the intense fragmentation towards middle-lower Tibagi has provided a significant increase in inbreeding coefficient (F_{IS}), loss of alleles and reduced genetic diversity compared to populations from the upper Tibagi (Table 1). These factors also influenced the genetic differentiation between populations of the upper Tibagi (where fragmentation is less evident) and the others. The only exception to this result was observed in *A. sellowiana*. When analyzing the species *A. sellowiana* with microsatellite markers we also observed the formation of two distinct groups of populations (Dendrogram using Nei´s genetic distance, 1978), one comprised of populations from the middle Tibagi region and the other pertaining to lower Tibagi (Oliveira, 2011 submitted). Even though *A. sellowiana* is able to perform vegetative reproduction and also its propagules can reach as far as 10m distance from mother tree, we evidenced highly significant negative values of F_{IS} indicating excess of heterozygosity in tree of the five populations. When considering the effect of recent genetic bottlenecks in these populations, the values obtained by the software bottleneck indicated that four of the five populations showed significant values of heterozygosity excess, when considering the Infinite Allele Model, and only one population presented significant levels of heterozygosity excess for all tree models (Infinite Allele Model, Two-phase Model and Stepwise Mutation Model). Migration rates calculated for these populations demonstrated that exchange of 30% of migrants between two populations from the middle Tibagi region. Such results demonstrate that some of these populations are suffering more than others, the impact of fragmentation and also founder effect and gene flow are playing a key role in the shaping of the genetic variability within these populations of *A. sellowiana* promoting a weak balance between the evolutionary forces of genetic drift and migration in this post fragmentation period landscape.

Lately, this research group has been engaged in the development and inter-specific transference of microsatellite primers (SSR) for the species *L. divaricata* (Ruas et al., 2009), *A. sellowiana* (Ruas et al., 2010), *P. rigida*, *A. polyneuron*, *C. xanthocarpa*, *C. gonocarpum* and *A. polyneuron*. The use of SSR markers will allow more accurate estimates about the distribution of genetic variation and the effects of fragmentation and local adaptation between the species of the Tibagi River basin. Such is possible due to the codominant nature of these markers, allowing to access allelic information and thus to estimate the number of alleles, heterozygosity, inbreeding and gene flow among natural populations. Although much work is yet to be done with these and other species of the Tibagi River basin, the present studies certainly represent a great step for the comprehension of the present availability of genetic resources and of its relation to the life history of species from forests fragments pertaining to the Tibagi River basin.

7. Conclusion

In summary, the knowledge we have accumulated so far on the genetic structure of the studied populations allows us to infer that the large genetic differentiation that has been maintained between natural populations, especially among those of late successional stages, makes it imperative to conserve all forest remnants. One of the strategies that we believe can be effective for the conservation and restoration of these ecosystems is the establishment of green corridors to restore communication of pollinators and dispersers between forest fragments of the Tibagi River basin. Moreover, the development of SSR markers for tropical tree species would first increase our knowledge on those species genome and help the investigation of the genetic determination and evolution of some important adaptations, such as flooding tolerance. Second, it would advance our understanding of the effects of habitat fragmentation over the riparian forests' diversity.

8. Acknowledgment

The authors are grateful to Conselho Nacional de Pesquisa (CNPq), Fundação Araucária, Coordenação de Aperfeiçoamento de Pessoal de Nível Superior (CAPES), Universidade Estadual de Londrina (UEL) and Universidade Estadual do Norte do Paraná (UENP) for financial support.

9. References

Aguilar, R., Quesada, M., Ashworth, L., Herrerias-Diego, Y., & Lobo J. (2008). Genetic consequences of habitat fragmentation in plant populations: susceptible signals in plant traits and methodological approaches. Molecular Ecology, Vol. 17, No. 24, (December 2008), pp. 5177-5188, ISSN 1365-294X

Antonelli, A. & Sanmartín, I. (2011). Why there are so many plant species in the Neotropics? Taxon, Vol. 60, No.2, (April 2011), pp. 403 – 414, ISSN 0040-0262

Bailey-Serres, J. & Voesenek, L.A.C.J. (2008). Flooding Stress: Acclimations and Genetic Diversity. Annual Review of Plant Biology, Vol. 59, (December 2008), pp. 313-339, ISSN 1543-5008

Bawa, K.S. (1990). Plant-pollinator interactions in tropical rainforests. Annual Review of Ecology, Evolution, and Systematics, Vol. 21, No. 1, (November 1990), pp. 399-422, ISSN 1543-592X

Bekessy, S.A., Ennosb, R.A., Burgmana, M.A., Newtonc, A.C. & Adesd, P.K. (2003). Neutral DNA markers fail to detect genetic divergence in an ecologically important trait. Biological Conservation, Vol. 10, No. 2, (April 2003), pp. 267-275, ISSN 0006-3207

Brooks, T.M., Mittermeier, R.A., Mittermeier, C.G., da Fonseca, G.A.B., Rylands A.b., Konstant, W. R., Oldfield, S., Magin, G. & Hilton-Taylor, G. (2002). Habitat loss and extinction in the hotspots of biodiversity. Conservation Biology,16: p. 910–923.

Crawford, R.M.M. & Braendle, R. (1996). Oxygen deprivation stress in a changing enviroment. Journal of Experimental Botany, Vol. 47, No. 295, (February 1996), pp. 145-159, ISSN 1460-2431

Damasceno, J.O., Ruas, E.A., Rodrigues, L.A., Ruas, C.F., Bianchini, E., Pimenta, J.A. & Ruas, P.M. (2011). Genetic differentiation in Aspidosperma polyneuron (Apocynaceae) over a short geographic distance as assessed by AFLP markers. Genetics and Molecular Research, Vol. 10, No.2, (June 2011), pp. 1180-1187, ISSN 1676-5680

Dean, W. (1996). A ferro e Fogo – A História e a devastação da Mata Atlântica brasileira, Companhia das Letras, ISBN 85-7164-590-6, São Paulo, Brazil

Debinski, D. M. & Holt, R. D. (2000). A survey and overview of habitat fragmentation experiments.Conservation Biology 14: p. 342-355.

De Carvalho, M.C.C.G., Silva D.C.G., Ruas, P.M., Medri M.E., Ruas E.A. & Ruas, C.F. (2008). Flooding tolerance and genetic diversity in populations of Luehea divaricata. Biologia Plantarum, Vol.52, No.4, (December 2008), pp. 771-774, ISSN 1573-8264

Dias, A., Latrubesse, E. M., Galinkin, M. (2002). Projeto corredor ecológico Bananal Araguaia, Brasília, 120 p.

Dick, C.W., Hardy, O.J., Jones, F.A., & Petit, R.J. (2008). Spatial scales of pollen and seed-mediated gene flow in tropical rain forest trees. Tropical Plant Biology, Vol. 1, No. 1, (March 2008), pp. 20-33, ISSN 1935-9764

Drew, M.C. (1992). Soil aeration and plant root metabolism. Soil Science, Vol. 154, No.4, (October 1992), pp. 259-268, ISSN 1538-9243

Fundação SOS Mata Atlântica & Instituto Nacional de Pesquisas Espaciais (INPE). (2011). Atlas dos remanescentes florestais da mata atlântica. São Paulo, Brazil

Futuyma, D.J. (2005). Evolution. Sinauer, ISBN 0-87893-187-2 (hardcover), Sunderland.

Givnish, T.J. (1999). On the causes of gradients in tropical tree diversity. Journal of Ecology, Vol. 87, No. 2, (March 1999), pp. 193-210, ISSN 1365-2745

Govaerts, R. (2001). How many species of seed plants are there? Taxon, Vol. 50, No.4, (November, 2001), pp.1085–1090, ISSN 0040-0262

Hamrick, J. L., Linhart, Y. B. & Milton, J. B. (1979). Relationships between life history characteristics and electrophoretically-detectable genetic variation in plants. Ann. Rev. Ecology and Systematics. 10:173-200.

Hamrick, J. L. (1983). The distribution of genetic variation within and among natural plant populations. In: Schone-Wald-Cox, C. M., Chambers, S. H.; MacByde, B., Thomas, L. (Ed.). Genetics and Conservation. Menlo Park: Benjamin Cummings, 1983. p. 335-348.

Howe, H.F., & Smallwood, J. (1982). Ecology of seed dispersal. Annual Review of Ecology, Evolution, and Systematics, Vol. 13, No. 1, (November 1982), pp. 201–228, ISSN 1543-592X

Instituto Brasileiro de Geografia e estatística (IBGE). (2004). Mapa de Biomas do Brasil, Rio de Janeiro, Brazil

IBGE – Instituto Brasileiro de Geografia e Estatística. (2006). Manual técnico de uso da terra. Rio de Janeiro: IBGE, 2 ed. (Manuais técnicos em Geociências, n.7). 91 p.

Instituto Brasileiro de Geografia e Estatística (IBGE). (2007). Censo Populacional 2005, Rio de Janeiro, Brazil

Kageyama, P. Y. (1987). Conservação in situ de recursos genéticos de plantas. Instituto de Pesquisas e Estudos Florestais - IPEF, 35: p. 7-37.

Kala, C. P. (2000). Status and conservation of rare and endangered medicinal plants in the Indian trans-Himalaya. Biological Conservation, 93: p. 371-379.

Kawecki, T.J., & Ebert, D. (2004). Conceptual issues in local adaptation. Ecology Letters, Vol. 7, No. 12, (December 2004), pp. 1225-1241, ISSN 1461-023X

Leimu, R., & Fischer, M. (2008). A meta-analysis of local adaptation in plants. PLoS ONE, Vol. 3, No. 12, (December 2008), pp. e4010, ISSN 1932-6203

Lovejoy, T. E. (1983). Ecological dynamics of tropical forest fragments. In: Sutton, S. L.; Whitmore, T. C., Chadwick, A. C. Tropical Rain Forest: Ecology and Management. Oxford: Blackwell Scientific Publication, p. 377-384.

Loveless, M. D. & Hamrick, J. L. (1987). Ecological determinants of genetic structure in plant populations. Annual Review of Ecology and Systematics, 15: p. 65-95.

Lowe, A.J., Boshier, D., Ward, M., Bacles, C.F.E., & Navarro, C. (2005). Genetic resource impacts of habitat loss and degradation; reconciling empirical evidence and predicted theory for neotropical trees. Heredity, Vol. 95, No. 4, (August 2005), pp. 255-273, ISSN 13652-540

Matallana,G., Wendt, T., Araujo, D.S.D. & Scarano, F.R. (2005). High abundance of dioecious plants in a tropical coastal vegetation. American Journal of Botany, Vol. 92, No. 9, (September 2005), pp. 1513-1519, ISSN 1537-2197

Medeiros, R.; Irving, M. & Garay, I. (2004). A proteção da natureza no Brasil: Evolução e conflitos de um modelo em construção. Revista de desenvolvimento econômico, Vol. No.9, (January, 2004) ISSN 1516-1684

Medri, C., Ruas, E.A., Medri, M.E., Ruas, C.F., Sayhun, S.A., Medri, P.S., Silva, D.C.G., Bianchini, E. & Ruas, P.M. (2011). Genetic diversity and flooding survival in Aegiphila sellowiana (Lamiaceae), a typical tree species from upland riparian forests. Genetics and Molecular Research, Vol. 10, No. 2, (June 2011), pp. 1084-1091, ISSN 1676-5680

Medri C., Ruas E.A., Ruas C.F., Medri P.S., Medri M.E. & Ruas, P. M. (2011). Population genetic structure of the tropical tree species Aegiphila sellowiana (Lamiaceae). Genetic and Molecular Research (in press).

Medri, M.E., Bianchini, E., Shibatta, O.A. & Pimenta, J.A. (2002). A Bacia do Rio Tibagi, Câmara Brasileira do Livro, ISBN 85902390-1-2/85-902392-1-7/85-902394-1-1/85-902395-1-9, Londrina, Brazil

Naiman, R.J. & Decamps, H. (1997). The ecology of interfaces: riparian zones. Annual Review of Ecology and Systematics, Vol. 28, (November 1997), pp. 621-658, ISSN 0066-4162

Nei, M. (1973) Analysis of gene diversity subdivided populations. Proceedings of the National Academy of Science, USA 70: p. 3321-3323.

O'Gorman, E. (1972). The invention of America: An inquiry into the historical nature of the New World and the meaning of its history. Greenwood Press, ISBN 0837164702, Westport, USA

Oliveira, R.B.R. (2010). Estrutura genética de cinco populações de Aegipila sellwiana Cham. (Lamiaceae) da bacia do Rio Tibagi por marcadores microssatélites. Master degree dissertation, Universidade Estadual de Londrina, Londrina, Brazil, 63 p.

Ribeiro, M.C., Metzger, J.P., Martensen, A.C., Ponzoni, F.J. & Hirota, M.M. (2009). The Brazilian Atlantic Forest: How much is left, and how is the remaining forest distributed? Implications for conservation. Biological Conservation, vol. 142, No.6, (March 2009), p. 1141-1153, ISSN 0006-3207

Ruas, E.A., Conson, A.R.O., Costa, B.F., Damasceno, J.O., Rodrigues, L.A., Reck, M., Vieira, A.O.S., Ruas, P.M. & Ruas, C.F. (2009). Isolation and characterization of ten microsatellite loci for the tree species Luehea divaricata Mart. (Malvaceae) and intergeneric transferability. Conservation Genetics Resources, Vol.1, No.1, (August 2009), pp.245-248, ISSN 1877-7260

Ruas, E.A., Ruas, C.F., Medri, P.S., Medri, C., Medri, M.E., Bianchini, E., Pimenta, J.A., Rodrigues, L.A. & Ruas, P.M. (2011). Anatomy and genetic diversity of two populations of Schinus terebinthifolius (Anacardiaceae) from the Tibagi River basin in Paraná, Brazil. Genetics and Molecular Research, Vol. 10, No. 1, (March 2011), pp. 526-536, ISSN 1676-5680

Ruas, E.A., Damasceno, J.O., Conson, A.R.O., Costa, B.F., Rodrigues, L.A., Reck, M., Vieira, A.O.S., Ruas, C.F., Medri, C. & Ruas, P.M. (2011). Isolation and characterization of eleven polymorphic microsatellite loci in Aegiphila sellowiana and their transferability. Biologia Plantarum, Vol. 55, No. 2, (June 2011), pp. 396-399, ISSN 1573-8264

Sahyun, S.A., Ruas, E.A., Ruas, C.F., Medri, C., Souza, J.R.P., Johan, L.A.P.S., Miranda, L.V. & Ruas, P.M. (2010). Genetic Variability of three Natural Populations of Maytenus aquifolium (Celesteraceae) from Telêmaco Borba, Paraná, Brazil. Brazilian Archives of Biology and Technology, Vol. 53, No. 5, (September/October 2010), pp.1037-1042, ISSN 1516-8913

Savolainen, O., Pyhäajärvi, T., & Knürr, T. (2007). Gene flow and local adaptation in trees. Annual Review of Ecology, Evolution, and Systematics, Vol. 38, No. 1, (December 2007), pp. 595-619, ISSN 1543-592X

Schultz, J. (2005). The ecozones of the world: The ecological divisions of the geosphere. Springer, ISBN 978-3-540-20014-7, New York, USA

Shannon, C.E. & Weaver, W. (1949) The mathematical theory of communication. Urbana, The Univeristy of Illinois Press 117pp.

Silva, D.C.G., Carvalho, M.C.C.G., Ruas, P.M., Ruas, C.F. & Medri, M.E. (2010). Evidence of ecotypic differentiation between populations of the tree species Parapiptadenia rigida due to flooding. Genetics and Molecular Research, Vol. 9, No. 2, (May 2010), pp. 797-810, ISSN 1676-5680

Souza L. B. 2011. Estrutura genética de populações de Parapiptadenia rigida (Benth) Brenan (Leguminosae-Momosideae) na região sul do Brasil por marcadores AFLP. Master degree Dissertation, Universidade Estadual de Londrina, Londrina, Brazil, 57p.

Stöcklin, J., Kuss, P. & Pluess, A.R. (2009). Genetic diversity, phenotypic variation and local adaptation in the alpine landscape: case studies with alpine plant species. Botanica Helvetica, Vol. 119, (November 2009), pp.125–133, ISSN 0253-1453

Takhtajan, A. 1986. Floristic regions of the world. University of California Press [An English translation of the original Russian book published in 1978], ISBN 0520040279 London and Berkeley, USA

Thomas, W. W. (1999). Conservation and monographic research on the flora of Tropical America. Biodiversity and Conservation, Vol. 8, No. 8, (August 1999), pp. 1007–1015, ISSN 0960-3115

Torezan, J.M.D. & Silveira, M. (2002). Fatores ambientais, diversidade e similaridade em florestas da bacia do rio Tibagi, In: A Bacia do rio Tibagi, Medri, M.E., Bianchini, E., Shibatta, O.A. and Pimenta, J.A, pp. 125-131, Câmara Brasileira do Livro ,ISBN 85902390-1-2/85-902392-1-7/85-902394-1-1/85-902395-1-9, Londrina, Brazil

Wittmann, A.O., Piedade, M.T.F., Wittmann, F., Schöngart, J. & Parolin, P. (2007). Patterns of structure and seedling diversity along a flooding and successional gradient in amazonian floodplain forests. Pesquisas, série Botânica, No. 58, pp. 119-138, ISSN 0373-840X

Wright, S. (1951). The genetical structure of populations. Annals of Eugenics 15: p. 323-354.

Young, A., Boyle, T., & Brown, A.H.D. (1996). The population genetic consequences of habitat fragmentation for plants. Trends in Ecology & Evolution, Vo. 11, No. 10, (October 1996), p. 413-418, ISSN 01695-347

6

Living on the Edge: Various Modes of Persistence at the Range Margins of Some Far Eastern Species

Elena Artyukova, Marina Kozyrenko, Olga Koren,
Alla Kholina, Olga Nakonechnaya and Yuri Zhuravlev
Institute of Biology and Soil Science, Far East Branch of Russian Academy of Sciences
Russia

1. Introduction

Present-day patterns of plant distribution have been formed under the influence of various biotic and abiotic factors. Plant distribution reflects the habitat preferences of species and the outcome of their competition as well as the complex evolutionary processes resulting in the specificity of mating systems, the genetic structure of different species and other aspects of species biology. Together, these factors determine the current ranges and distributions of plant species. At the edge of a species' range, the significance of particular interactions with the environment becomes more pronounced. However, our understanding of this class of interactions is limited. There is debate about whether these interactions represent a distinct and ordered set of related phenomena or whether they are unrelated and without order. Different approaches to this problem are needed in different situations. Understanding the processes of microevolution in species at the edges of their ranges is of great interest, particularly in view of the continuing decline in worldwide biodiversity and ongoing and future climate changes. When the area of a plant's habitat is sufficient, most populations exist in a relatively stable condition, and changes in their genetic structure follow slow processes, such as gene flow and genetic drift. However, in populations growing at the edge of their range, the rates of genetic processes can change dramatically. At the limits of the climatic and ecological tolerance of species, populations usually become smaller and more fragmented. These populations are generally less genetically diverse than those living at the center of the range because they exist in less favorable habitats and at lower densities, and consequently, they may be more prone to extirpation (Hampe & Petit, 2005; Vucetich & Waite, 2003). However, some species may have existed as groups of isolated populations for thousands of generations. The long-term survival and evolution of a species depends on the maintenance of sufficient genetic variability within and among populations. Patterns of population genetic diversity have been shown to be generally shaped by past climate-driven range dynamics, rather than solely by stochastic demographic and genetic processes (Hewitt, 2004). Given the enormous variety of plant life forms and their habitats as well as the complexity of their evolutionary histories, it is difficult to accept as a general rule that all marginal populations will exhibit lower genetic diversity than those from the center of a species' range. Moreover, this rule would not hold true for all rare plants, which are associated with different causes underlying their rarity. A comparison of rare species belonging to different plant families and with different evolutionary histories may aid in inferring a number of scenarios under which a species may persist at the edge of its range. These scenarios may include common principles that do not correspond well to the "center-periphery" hypothesis.

2. The region and the species selected

The southern part of East Asia together with tropical Asia, is considered one of the centers of the origin and diversification of many plant taxa. The southern region of the Russian Far East (Primorsky Territory, Primorye) is located at the eastern edge of the Asian continent. The majority of this region is mountainous, with the Sikhote Alin Mountains extending throughout most of the area. The geographic location of the region accounts for the variety of its flora: it includes mountainous tundra areas, coniferous forests and coniferous-deciduous forests, and part of the lowlands surrounding Khanka Lake is occupied by forest-steppe. The flora of the region is unique and is characterized by a very complex mix of representatives of different kinds of vegetation. Unlike many other regions at the same latitude, most of this area was not glaciated during the Pleistocene; glaciers were limited to the highest peaks of Sikhote Alin, and the vegetation has undergone uninterrupted development since the Pliocene. The modern species complexes of Sikhote Alin have been formed during numerous migration processes under the influence of global climate change, the specific impact of the region's proximity to the ocean and marine transgressions. The region maintains a large number of rare and endangered species that originated in earlier epochs. Unfortunately, many of the previously abundant plants of the area have already become rare and are disappearing as a result of increasing anthropogenic pressure. The removal of rare and disappearing plants from their native habitats has led to the disruption of natural ecosystems and has significantly impoverished the biodiversity of the region.

Over 2,500 vascular plant species are represented in the flora of the southern part of the Russian Far East, and more than 340 species are listed in the Red Data Book of Primorsky Krai (2008) as endemic, rare or endangered. A large number of rare species are endemic, restricted to certain habitats, or their northern distribution limits are located in the region. Some of these species are relics of Tertiary flora with extremely limited ranges, whereas others exhibit range habitat preferences occupying tiny areas at specific locations. We have chosen a number of rare Far Eastern species (Figure) characterized by different life history traits for the present study. All of these species are listed in the Rare Plant Species of the Soviet Far East and Their Conservation (Kharkevich & Kachura, 1981), Red Data Book of Primorsky Krai (2008) and the Red Data Book of the Russian Federation (1988). The genetic diversity and population structure of each of these species have been studied using the following markers: allozymes, dominant DNA markers (random amplified polymorphic DNA, RAPD, and/or inter simple sequence repeat, ISSR) and sequences of noncoding regions of the chloroplast genome (cpDNA). Some life history traits and the main parameters related to genetic diversity and population structure in these species are presented in Table. Below, we provide a detailed description each of these species.

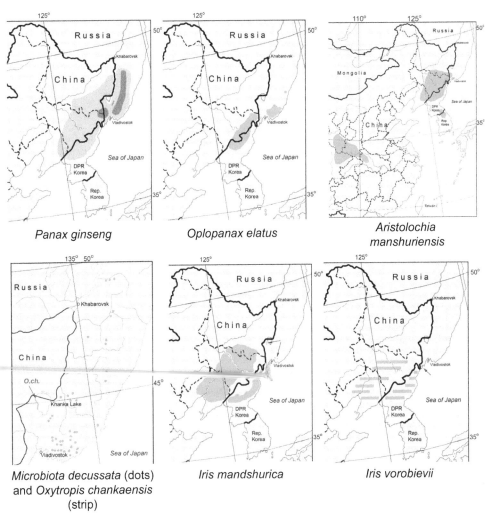

Fig. 1. Geographic ranges of rare Far Eastern species under study (according to Alexeeva, 2008; Huang et al., 2003; Kharkevich & Kachura, 1981; Shu, 2000; Xiang & Lowry, 2007; Zhu et al., 2010).

2.1 Ginseng (oriental ginseng), *Panax ginseng* C.A. Meyer

Ginseng, *Panax ginseng* C.A. Meyer, is a representative of the *Panax* L. genus related to an ancient family of angiosperms, the Araliaceae, members of which have been found in Cretaceous deposits. Most members of the Araliaceae are distributed in tropical and subtropical regions, with some species reaching the temperate zone. The *Panax* genus consists exclusively of herbaceous perennial plants, and all species in this genus are distinguished by the peculiar structure of an aboveground shoot that dies annually, whereas most members of the Araliaceae are trees or shrubs (Grushwitsky, 1961). The uniqueness of

Species*	Life-hystory traits					
	Life form	Life-span (years)	Clonality	Mating system	Pollination	Seed dispersal
Panax ginseng, E	PH	140	NC	S, A	I	G, A
Oplopanax elatus, V	S	300	C	S	I	G, A
Aristolochia manshuriensis, E	WL	50	NC	S	I	W, Wa
Microbiota decussata, V	S	250	C	S	W	G
Oxytropis chankaensis, V	PH	50	NC	S	I	W, Wa, G
Iris vorobievii, E	PH	≤7	PC	S	I	G, A
Iris mandshurica, V	PH	25	C	S	I	G, A

Species*	Genetic diversity parameters at population level								
	Allozymes			Dominant markers			Chloroplast DNA		
	P_{95}	He	F_{ST}	P_{95}	He	G_{ST}	$\pi \cdot 10^{-3}$	h	G_{ST}
Panax ginseng, E	7.6	0.022	0.204	4.0	0.013	0.249	–	–	–
Oplopanax elatus, V	25.0	0.113	–	23.5	0.088	0.293	–	–	–
Aristolochia manshuriensis, E	25.0	0.108	0.065	36.2	0.141	0.112	–	–	–
Microbiota decussata, V	–	–	–	45.1	0.249	0.352	0.603	0.954	0.090
Oxytropis chankaensis, V	37.1	0.294	0.025	66.9	0.290	0.135	0.480	0.703	0.146
Iris vorobievii, E	–	–	–	32.5	0.104	–	0.587	0.912	–
Iris mandshurica, V	–	–	–	31.3	0.108	–	0.285	0.733	–
Mean values for species with restricted ranges**	29.9	0.095	0.206	–	0.280	0.210	–	–	0.637a 0.165g

Table 1. Life-history traits and parameters of genetic variation in populations of rare Far Eastern species studied. *, Categories of rarity are given with the species name according to the Red Data Book of Primorsky Krai (2008): E, endangered; V, vulnerable. Life form: S, shrub; WL, woody liana; PH, perennial herb. Clonality: C, clonal; PC, poor clonal; NC, non-clonal. Mating system: S, sexual; A, apomixis. Pollination: W, wind; I, insect. Seed dispersal: W, wind; Wa, water; A, animal; G, gravity. Genetic diversity parameters at population level: P_{95}, percentage of polymorphic loci (95% criterion); He, expected genetic diversity; π, nucleotide diversity; h, haplotype diversity; F_{ST}, and G_{ST}, indices of genetic differentiation among populations; –, not determined. **, Mean values of genetic variation parameters in species with restricted ranges are cited from Gitzendanner & Soltis, 2000 (for allozyme data) and Nybom, 2004 (for dominant DNA marker data); G_{ST} based on chloroplast DNA data are cited from Petit et al. (2005): a, for angiosperms; g, for gymnosperms.

The genus *Panax* makes it difficult to ascertain its alliances with other genera of Araliaceae. The range of *Panax* is divided by the Pacific Ocean into two parts: an East Asian and a North American region (Grushwitsky, 1961). The intrageneric systematics of *Panax* has been revised as a result of the similarity of different *Panax* species habitus and the presence of intermediate forms that complicate species discrimination. Most *Panax* species are found in Eastern and Southeastern Asia, with the exception of *P. quinquefolium* L. and *P. trifolius* L., which inhabit North America. *P. ginseng* and *P. quinquefolium* have a chromosome number of 2n = 48, whereas most other *Panax* species exhibit 2n = 24.

In the past, *P. ginseng* was distributed throughout a wide territory in the forests of the Far East. It was estimated in the 1950s that in the first half of the 20th century, wild-growing ginseng plants could be found in large forestlands from 40° to 48° northern latitude and from 125° to 137° eastern longitude, covering approximately 500,000 square kilometers (Grushwitsky, 1961). At that time, the ginseng populations growing in Russia represented the northeastern boundary of the species' geographic range, whereas its main habitat area was located south and west of this territory, covering Heilongjiang, Jilin and Liaoning Provinces in China and the northern part of Korea. The natural range of ginseng is now drastically reduced (Figure). Wild-growing ginseng had disappeared completely from Korea and Liaoning Province in China by the 1930s and almost completely from Heilongjiang and Jilin Provinces in China by the 1990s (Zhuravlev & Kolyada, 1996). The distribution of ginseng has currently narrowed to a few patches in Russia and China (Zhuravlev & Kolyada, 1996; Zhuravlev et al., 2008). The largest patch of this species is located in the southern part of the Sikhote Alin mountain range; another population inhabits the southwest region of Primorye and Jilin and Heilongjiang Provinces (Changbai Mountains) in China; and a third population is located in the western part of Primorsky Territory (Figure, dark green). Reports of wild ginseng in China have become rare. In Russia, ginseng has been listed in the Red Book since 1975 as a federally threatened species (Red Data Book of the Russian Federation, 1988), and at present, Primorsky Territory in Russia is the only place in the world where natural ginseng populations exist, representing a remnant gene pool of wild-growing ginseng.

Ginseng is an herbaceous perennial species with an annually moribund shoot. It grows in certain special habitats and persists only in remote locations. It is very difficult to estimate the actual abundance of wild-growing ginseng in nature because of its ability to undergo lengthy dormancy, which may last from one to several dozen years; therefore, the recorded number of vegetative ginseng plants does not correspond to their true abundance. I.V. Grushwitsky investigated natural ginseng habitats in the 1950s and wrote that this plant did not exhibit a tendency toward extinction for the most part, despite its relic origin and low rate of regeneration (Grushwitsky, 1961). However, he also noted that human activity has a destructive influence on natural ginseng populations, reducing them to potentially threatened levels. In addition, logging, fires and the shallowing of rivers hamper the restoration of ginseng populations. Poaching and overexploitation of ginseng resources appear to be the most important and most obvious reasons for the reduction and exhaustion of ginseng populations today (Zhuravlev & Kolyada, 1996).

Ginseng reproduces exclusively through seeds and is nearly incapable of vegetatively reproducing (Grushwitsky, 1961; Zhuravlev & Kolyada, 1996). This species is characterized by a mixed mating system and the ability to produce seeds via autogamy, outcrossing or agamospermy without pollination (Koren et al., 1998). Presumably, self-pollination has prevailed in natural ginseng populations because of their low plant density. The occurrence of outcrossing, which is carried out by insect pollen transfer, cannot be excluded in natural environments, but this appears to be very rare and likely does not play a significant role in ginseng pollination. *P. ginseng* has also been shown to be a facultative apomict with a type of agamospermy resembling diplospory (Koren et al., 1998; Zhuravlev et al., 2008).

Studies on the genetic variability of *P. ginseng* using different molecular markers (allozymes, RAPD, ISSR) have detected very low levels of genetic polymorphism (Koren et al., 2003; Zhuravlev et al., 2008; Reunova et al., 2010a). Only 3 allozyme loci among 39 studied and only one of 74 RAPD loci were found to be polymorphic. These low levels (Table) do not differ significantly from estimates obtained with ISSR markers (P_{ISSR} = 9.3%, He_{ISSR} = 0.014, Reunova et al., 2010a). These data are also in agreement with the results of genetic studies on cultivated ginseng sampled from China and Korea (e.g., Kim & Choi, 2003). Thus, *P. ginseng* is characterized by a lower level of genetic variation than the average values found for rare endemic species (Table).

An analysis of population subdivisions based on allozymes and dominant DNA markers showed a low level of differentiation of natural *P. ginseng* populations and a decrease in the total genetic diversity in the Sikhote Alin population (Zhuravlev et al., 2008). The great majority (up to 95%) of the genetic variability in this species is concentrated within populations, whereas only 4.1% of the total variation was found to be distributed among 8 sub-populations (estimated by allozymes, Zhuravlev et al., 2008). Approximately 25% of the variation is distributed between populations (estimated with DNA markers, Reunova et al., 2010a). No correlation between geographic and genetic distances was found for the investigated populations using allozyme and DNA markers.

Extremely low levels of genetic variation are usually found in endemic or relic plant species with narrow ranges which can often be connected to a species' life history and/or evolutionary events such as selection or genetic drift. The low genetic diversity of *P. ginseng* populations indicates that this species has experienced a severe genetic bottleneck. In particular, the lack of variation in the Sikhote Alin population may be a result of a founder effect because of the lack of a refugium during the last Pleistocene-Holocene cooling. This hypothesis is supported by the results of an analysis of the genetic relationships of extant ginseng populations on the basis of allozymes and DNA markers (Zhuravlev et al., 2008; Zhuravlev et al, 2010; Reunova et al., 2010a). Without other available natural populations to analyze, we can only assume that the center of the genetic richness and, possibly, the center of the origin of ginseng was located southwest of its extant natural populations, potentially in a place where there is great industrial activity in modern China.

Another reason for the low genetic diversity found in *P. ginseng* may be adaptive selection that has occurred in response to climate cooling and during the expansion of the species northward from its southern refugia. Some peculiarities of ginseng biology (such as its underdeveloped embryos and aboveground germination) indicate that this species evolved during a warm climate as a representative of the ancient thermophilic flora. A number of *P. ginseng* traits, especially related to its mating system, indicate recent adaptations (Koren et al., 1998; Zhuravlev et al., 2008).

Past evolutionary events have resulted in extant ginseng populations being characterized by very low genetic variation. However, this low genetic diversity did not prevent the species from surviving across a wide territory until recently. Despite its weak competitiveness, *P. ginseng* possesses some degree of ecological flexibility, certain adaptations to unfavorable environments and the ability to ensure seed reproduction via different pathways. Among these properties, increased individual longevity may be the main mechanism underlying the long-term survival of the species under unstable conditions. Discoveries of 100-year-old ginseng plants are occasionally recorded, and cases of 300-year-old plants are well known

(Zhuravlev & Kolyada, 1996). In addition, the life spans of individual plants may lengthen in accordance with their ability to undergo long-term dormancy (in the form of so-called "dormant roots"). Because ginseng maintains a generative stage starting from an age of 3-5 years and lasting until death, annually producing up to 100 seeds per plant, a single individual can maintain a population's size for hundreds of years, even if only a small proportion of its offspring survives and reaches generative age. Moreover, the seeds of one generation can germinate over a period of several years because to mature, the underdeveloped embryo requires an alternation of warm and cold periods that can take from one to several years.

At the same time, limited opportunities for recombination (autogamy, agamospermy) can result in further reducing the genetic diversity of ginseng. Moreover, because all three extant ginseng populations occupy the northeastern margin of its former distribution area, their gene pools are not enriched by gene flow from extirpated central populations. With the continuing reduction of natural ginseng populations as a result of human activity, this species may become extinct in the wild very soon.

2.2 Japanese devil's-club, *Oplopanax elatus* (Nakai) Nakai

Oplopanax elatus is another member of the ancient family Araliaceae. This species is a deciduous shrub with a spiny stem approximately 1 m in height and large palmately compound leaves. The species' distribution area is quite limited. In Russia, *O. elatus* occurs in the southern Primorye, where its range is represented by several isolated populations associated with the main mountain peaks of the southern Sikhote Alin. Outside Russia, *O. elatus* grows on the northern Korean Peninsula (Kurentsova, 1968; Zhuravlev & Kolyada, 1996) and in Jilin Province in China (Xiang & Lowry, 2007) (Figure). *O. elatus* is a valuable medicinal plant; its effects are similar to those of ginseng, and it is authorized for medical use (Kurentsova, 1968; Zhuravlev & Kolyada, 1996; Schreter, 1975). This species is particularly vulnerable as a result of intensive harvesting and habitat disturbance due to fires and logging.

Oplopanax Miq. is a small genus that includes three species and demonstrates a classical Eastern Asian and North American disjunct distributional pattern. *O. horridus* occurs on the Pacific coast of the United States and Canada as well as around Lake Superior. *O. japonicus* is endemic to the Japanese Islands (Hokkaido, Honshu and Shikoku). This disjunct distribution pattern is observed for many plant genera and is explained by the existence of the land bridges between Eurasia and North America and between the Japanese Archipelago and the mainland during the Pliocene-Pleistocene. Based on the internal transcribed spacer sequences of nuclear ribosomal DNA (ITS rDNA), phylogenetic analysis confirmed the origin of the three *Oplopanax* species from a common East Asian ancestor. The closely related species *O. horridus* and *O. elatus* form a sister pair, and both are closely related to *O. japonicus* (Artyukova et al., 2005). An ancestral form that gave rise to *O. horridus* and *O. elatus* could have survived in the coastal zone during global cooling and the strengthening of climate continentality during the Early Pliocene and subsequently spread along the Pacific coast and across the Bering Bridge into North America.

In southern Primorye, *O. elatus* is mainly confined to the orotemperate belt in the altitudinal range of ca. 800 to 1,500 m above sea level and is a common species in the understory of fir-spruce forests (Kurentsova, 1968) occurring in moist, well-drained ecosystems. This species, which is clearly representative of the thermophilic and hydrophilous Turgai flora, has found specific refugia in the understory of fir-spruce forests, which favors its growth. Only under these conditions does the species occur at lower altitudes in coniferous–deciduous forests. However, its growth in mixed forests is hindered not only by low humidity, but also by competition from shrubs and grasses, which are abundant in the understory of mixed forests in Primorye. In spruce forests, the shrub and herb layers are poorly developed, and *O. elatus* has an opportunity to achieve a wider distribution (Kurentsova, 1968). In these habitats, the species is usually very abundant and sustainable. However, it is unable to withstand competition from heliophilous species, which rapidly fill gaps after disturbances to primary stands, such as natural or artificial fires or logging.

O. elatus mostly reproduces vegetatively through lateral branches that form adventitious roots when in contact with the soil. This species grows clonally, and individual clones can include 20 or more shoots, which can remain connected by decumbent stems to the parental plant for extended periods of time (Kurentsova, 1968). The life span of a single shoot is up to 40 years, and the overall life span of an individual from germination to the death of all parts of the clone can reach 300 years (Zhuravlev & Kolyada, 1996). Propagation of *O. elatus* through seeds is impeded; the seed set on individual plants is high, but most seeds exhibit underdeveloped embryos. Delayed embryo development is an ancient feature of many Araliaceae. A small number of *Oplopanax* seeds germinate in the second year, and most of the seedlings die.

An analysis of genetic diversity using dominant molecular DNA markers (RAPD and ISSR) allowed the levels of intra- and interpopulation variability to be evaluated in three geographically isolated *O. elatus* populations (Reunova et al., 2010b). The level of genetic diversity in *O. elatus* is much less (Table) than that in *Kalopanax septemlobus* (P = 59.2%, He = 0.119; Huh et al., 2005) or *Dendropanax arboreus* (P = 70.2%, He = 0.253; Figueroa-Esquivel et al., 2010), which are Araliaceae species with more continuous ranges that have been studied using dominant markers.

The *O. elatus* sample from Mt. Litovka (Livadiiskii Range, southern Sikhote Alin) with the highest level of RAPD variability (P = 29.41%, He = 0.110; Reunova et al., 2010b) is characterized by a level of genetic diversity similar to that detected based on allozyme data (P = 25.0%, Ho = 0.131, He = 0.113; Kholina et al., 2010). Allozyme analysis has shown a slight excess of heterozygotes in this population. This may be attributable in part to vegetative reproduction of pre-existing heterozygous genotypes and/or to selection favoring heterozygotes that are better able to adapt to new environments. The low polymorphism and allelic diversity in *O. elatus,* along with a rather high level of observed heterozygosity (Ho), may be indirect evidence of the effect of genetic drift, which reduces allelic diversity and, consequently, the proportion of polymorphic loci. It seems likely that the source from which the studied *O. elatus* populations were established was heterozygous plants, especially if they were characterized by increased viability. In addition, an excess of heterozygotes and high values of Ho may be explained by mutations that could have arisen in long-lived clones and been maintained by vegetative reproduction (similar to the serpentine endemic *Calystegia collina*; Wolf et al., 2000). Given the mainly vegetative mode of reproduction of *O. elatus* and the total life span of an individual accession lasting up to 300 years, accumulation of mutations maintained by vegetative reproduction cannot be ruled out.

Thus, based on different nuclear DNA marker data, the levels of genetic variation in the populations located near the northern edge of the *O. elatus* range appear to be comparable with the mean values reported for rare plant species (Table). In addition to its small distribution range and ecological specificity, the low level of genetic variation detected in *O. elatus* may be determined by the species' history. Reductions in polymorphism levels are caused by population bottlenecks resulting from dramatic decreases in population size. Bottleneck events have been proven to be the most probable reason for low polymorphism levels in relic Araliaceae species such as *P. ginseng* (Koren et al., 2003) and *Dendropanax morbifera* (Kim et al., 2006). As mentioned above, *O. elatus* inhabited deciduous Turgai forests. During periods of glaciation, broadleaved forests experienced dramatic range contractions and were forced to retreat southward, and the zone of high mountain vegetation was shifted to lower elevations. It has been suggested that coastal areas and the temporarily emerged continental shelves of the Sea of Japan connecting the Japanese Archipelago to the continent may have served as refugia for coniferous and broadleaved mixed forests (Sakaguchi et al., 2010). During its isolation in refugia and its subsequent expansion, *O. elatus* might have adapted to life in the understory of the fir-spruce forests. The possibility cannot be excluded that contractions in the species' population size following re-establishment from just a few founders could have happened more than once and may be occurring today, when these processes are associated with intense human activity (logging, fires and plant harvesting).

An analysis of the clonal structure of the population from Mt. Litovka indicated that in a sample of 29 accessions, there were a total of 22 multilocus genotypes. Only four of these genotypes were found more than once (from two to four times). The genotypic diversity (G/N) was determined to be 0.76, indicating that 76% of the individuals exhibited unique genotypes. Simpson's diversity index, D, is equal to 0.97, whereas the mean D value for 21 species of clonal plants has been 0.62 (Ellstrand & Roose, 1987). The existence of a great number of different genotypes along with low genetic variation has been reported for a number of species (Watkinson & Powell, 1993; Xie et al., 2005 and references therein). This phenomenon may be explained by the sporadic seed propagation in these species. In general, even a small number of individuals resulting from sexual reproduction is sufficient to make a population genotypically variable. Our findings imply that in the population of *O. elatus*, sexual reproduction succeeds periodically and results in the maintenance of a certain level of genotypic diversity. In addition, it has been suggested that events such as a presumptive origin from heterozygous founders, the accumulation of mutations and the retention of the changes via vegetative reproduction could promote the high level of genotypic variation observed.

The rarity of *O. elatus* is largely a result of this species' ecological specificity, dependence on humidity and low competitive ability. Anthropogenic influences (plant harvesting and habitat destruction) threaten the existence of this species. The flexibility of its reproductive system, combining different modes of reproduction, allows this species to renew heterozygous genotypes by clonal growth and to contribute additional variability resources through sporadic seed reproduction. At the same time, some features of the species' biology (the long life span of a single clone, overlapping generations and the ability to cross-pollinate) also help to maintain a certain level of polymorphism. The existing level of genetic diversity can be crucial to preventing the negative consequences associated with a small number of isolated populations and genetic drift.

2.3 Manchurian birthwort, *Aristolochia manshuriensis* Komarov

Manchurian birthwort, *Aristolochia manshuriensis* Kom., is a relic woody liana that belongs to the ancient angiosperm family Aristolochiaceae. This species is endemic to the Manchurian floristic region (Kitagawa, 1979) and occurs in the montane mixed forests of China and Korea (Figure). In the southern part of the Russian Far East, the species reaches the northern boundary of its range. *A. manshuriensis* prefers specific habitats in river floodplains at a certain altitude above sea level, along chutes and in steep slope foots, especially those with northern slope aspects, avoiding sunlit habitats associated with soil overheating. All Russian *A. manshuriensis* populations are located in the valleys of just three rivers and their tributaries. These populations are fragmented and separated by ridges. Within populations, plants grow at an uneven density, forming patches separated by 0.5–4 km from each other. The growth and expansion of natural *A. manshuriensis* populations are suppressed, and their natural regeneration is very poor (Kurentsova, 1968). At least three of four extant *A. manshuriensis* populations are now located in disturbed habitats and are experiencing strong anthropogenic pressure.

A. manshuriensis exhibits no means of vegetative reproduction, possesses a poor rooting ability and requires a long period of root formation (Shulgina, 1955). Seed reproduction in this species is usually successful, but spontaneous fruit set of only 2% has been documented. The seed set on a fruit is rather high because all mature fruits contain approximately 100 fully viable seeds. A histological analysis showed normal development of flower structures, a high level of pollen-grain fertility (97%) and a large number of pollen grains in the anthers of this species (Nakonechnaya et al., 2005).

Its seeds are adapted to spread by water and by wind (Nechaev & Nakonechnaya, 2009). These seeds are characterized by underdeveloped embryos, but seed germination is usually successful after winter dormancy or brief artificial stratification (Adams et al., 2005). Under favorable conditions, seedlings develop to the generative stage over 10-12 years, reaching a height of 15 m or more (with support) by this age. A long life span for a single *A. manshuriensis* plant is estimated to be more than 40 years. Under unfavorable conditions, plants can be kept in a virgin state for many years without flowering.

Similar to most Aristolochiaceae species, *A. manshuriensis* is an evolutionarily outcrossing plant exhibiting special adaptations of its flower structures (gynostemium) to ensure cross-pollination by insects. Autogamy and geitonogamy are possible, but pollinators are required to perform self-pollination as well as cross-pollination (Nakonechnaya et al, 2008). This species is characterized by long-term and abundant blossoming as well as a long life span of individual flowers, which increases the possibility of cross- and self-pollination. However, the presence of suitable insect pollinators during *A. manshuriensis* blooming is a precondition for successful seed reproduction in this plant. The flower structure of *A. manshuriensis* is matched to insect pollinators with a certain body size and thorax structure to allow the transfer of a sufficient number of pollen grains to the stigma. Among the visitors to *A. manshuriensis* flowers, flies of the genus *Pegoplata* (Anthomyiidae) are the only possible pollinators (Nakonechnaya et al., 2008). These flies have many other substrates available for breeding and larval feeding, which may be a cause of their rare visitations to *A. manshuriensis* flowers and, as a result, the rare fertilization and low fruit set of this species. Given that many Aristolochiaceae species are associated with specific insect pollinators (Nakonechnaya et al., 2008 and references therein), it may be assumed specific pollinators of marginal populations of *A. manshuriensis* have been lost as a result of shifts in the areas occupied by flora and fauna during the Pleistocene cooling. However, the high seed productivity of each mature fruit indicates the high potential fecundity of the species.

The mean level of genetic variability in *A. manshuriensis* is rather moderate based on allozymes (Nakonechnaya et al., 2007) and RAPDs (Table). The relatively high level of heterozygosity in *A. manshuriensis* populations compared with the average values for rare endemic species (Table) may be explained by outcrossing. Indeed, an excess of heterozygotes has been revealed within *A. manshuriensis* populations, despite their small population sizes, and no inbreeding has been found at the species level (Koren et al., 2009).

The relatively low level of population differentiation identified with molecular markers may be indicative of gene flow among populations in the present or recent past. Another possible reason for the low differentiation observed may be the greater degree of habitat integrity and the closer connections between populations that existed in the past (Koren et al., 2009). This explanation is supported by the low mean genetic distances between populations (D_N = 0.0096) and the value of gene flow between populations (Nm = 3.96) calculated based on G_{ST}, which shows that the isolation of the populations studied is incomplete.

Within two of the most disturbed populations, a statistically significant subdivision is shown with both allozyme and RAPD markers. Moreover, an effect of a genetic bottleneck is revealed in three populations that have undergone strong anthropogenic pressure (Koren et al., 2009). These findings indicate a reduction in gene flow between subpopulations in recent years as result of the intensive disturbance of their habitats.

Thus, the relic populations of *A. manshuriensis* exhibit moderate levels of genetic diversity. The low level of differentiation between the populations is probably connected with the evolutionary history of the species and the interaction of various factors, such as migration and selection. Despite the fact that all of the extant *A. manshuriensis* populations in the Russian part of the range are isolated from one another by ridges, they all grow in the valleys of rivers with their headwaters in the Borisovskoe Plateau and may have a common origin. It is likely that fragmentation of the *A. manshuriensis* populations occurred recently during the increase of human activities and the forest destruction that took place in the 20th century (Koren et al., 2009). The ongoing degradation of natural populations seems not to be associated with depletion of the gene pool and inbreeding depression. Because of the small size and fragmentation of these populations and their isolation from each other and from the species distribution center, genetic drift makes a significant contribution to the decrease in this species' genetic variability. In addition, anthropogenic influences, such as fires and uncontrolled harvesting of plants for their medicinal value (Bulgakov & Zhuravlev, 1989), also play a role in the contraction of the *A. manshuriensis* population size. Seed reproduction through outcrossing seems to be the only means for Manchurian birthwort to maintain a sufficient level of genetic variation. The preference of this species for specific ecotopes, its poor potential to undergo vegetative reproduction and its rare fertilization as well as its weak competitive ability in the virgin stage do not allow it to expand beyond the boundaries of its existing stands.

2.4 Russian arborvitae (Siberian Cypress, Russian Cypress), *Microbiota decussata* Komarov

The perennial evergreen decumbent coniferous shrub *Microbiota decussata* Kom. (Cupressaceae) is known from the southern part of the Russian Far East, where it occurs on some mountain peaks in the Sikhote Alin Mountains. The species is strongly restricted in its distribution, with its natural range stretching from 43° 00' N to 48° 50' N, from the subalpine zone of southeastern Primorsky Territory to the mountains in the Anyuy River basin in Khabarovsk Territory (Figure).

M. decussata is the sole species in the genus *Microbiota*, which is the only Cupressaceae genus endemic to the Sikhote Alin Mountains, and it is considered to represent one of the plant species that existed in this mountain system before the mountain-valley glaciers developed (Shlotgauer, 2011). The Sikhote Alin palynofloras contain pollen of Cupressaceae (Pavlyutkin et al., 2005 and references therein), but the cupressaceous pollen grains could not be distinguished beyond the family level. The fossil species *Cupressinoxylon microbiotoides* Blokhina from the Eocene/Oligocene deposits of Yuri Island of the Kurils is considered to have represented a putative species of the genus *Microbiota* (Blokhina, 1988). Of the extant Cupressaceae species, *M. decussata* is most closely related to *Platycladus orientalis* based on morphological data. The only fossil record of *M. decussata* was found in the Pavlovka lignite field (the southern Primorye) dated to the Pliocene or the late Eocene through late Oligocene (Pavlyutkin et al., 2005 and reference therein). In the Pavlovka deposits, wood of *M. decussata* was found together with fossilized wood of *P. orientalis* (Bondarenko, 2006). The presence of both species in the Pavlovka lignite field indicates their co-occurrence in the Oligocene/Pliocene plant associations of Sikhote Alin. Currently, *Platycladus orientalis* is a common tree in China and is widely cultivated elsewhere in Asia, eastward to Korea and Japan, southward to northern India and westward to northern Iran, whereas *M. decussata* is not found outside Russia. Its range is restricted to the Sikhote Alin, where the species exhibits a strongly disjointed distribution (Figure). Phylogenetic studies based on ITS rDNA and chloroplast markers (Gadek et al., 2000; Little et al., 2004) have confirmed the genetic affinity of the two taxa, which are placed in a sister clade to *Calocedrus*.

The heliophilous Russian arborvitae shrubs inhabit steep stony slopes and scree fields at altitudes from ca. 300 to 1,700 m above sea level (mainly at or above the timberline) under climatic conditions that impede the growth of forest vegetation. The xerophytic species *M. decussata* is a poor soil-tolerant plant that is resistant to a wide range of temperatures. As it grows under very harsh climatic and edaphic conditions, *M. decussata* is a pioneer in the colonization of cold stone deserts and participates in soil formation processes (Urusov, 1979). This low shrub with creeper and ascending branches often forms dense (crown density of 1.0) monodominant stands (Kurentzova, 1968; Krestov & Verkholat, 2003). In the southern part of its range, populations of *M. decussata* are more sustainable than those of another subalpine shrub, *Pinus pumila*. At the northern boundaries of its range, in the Anyuy and Chor River basins, the viability of *M. decussata* populations appears to be lower as a result of competitive inhibition by boreal species. Under these circumstances, fruiting of Russian arborvitae is rarely observed (Melnikova & Machinov, 2004), and the species is replaced by *P. pumila* (Kurentzova, 1968; Urusov, 1988).

M. decussata is an anemophilous monoecious plant that propagates sexually and by layering. The plant begins bearing fruits at 14–17 years, and the maximum life span of individuals has been determined to be approximately 250–300 years. The female cones of this species contain only a single 2 mm long naked seed. The seeds are able to disperse over a very short distance, dropping near the parent plant, though sporadic dispersal by animals cannot be ruled out. The seeds retain germination capacity for a long period but germinate almost exclusively after fires (Urusov, 1979). Outside the area occupied by their parent population, seedlings and juvenile plants are extremely rare (Kurentzova, 1968). *M. decussata* grows very slowly (3–7 cm a year) and can extend into new free habitats through the slow creeping of stems (up to 3–5 m in length) that produce roots at their nodes (Urusov, 1988).

M. decussata exhibits a high level of nuclear genome variation (Table), despite the restricted and fragmented range and geographical isolation of its populations (Artyukova & Kozyrenko, 2009). The gene diversity within populations from the southern and middle parts of the range of this species is slightly higher than that in the northern population from the Chor River basin, which is in close proximity to its northern range limits. The lower level of gene diversity in the northern population might be caused by severe temperature conditions interfering with sexual reproduction; thus, species dispersal occurs mainly by layering. The highest level of genetic diversity is found in the population from the central Sikhote Alin. The lack of population-specific RAPD markers and the similar high levels of diversity retained in the populations could be caused by a common gene pool and ancient polymorphism. The genetic differences among all pairs of populations (separated by 10–400 km) fit an isolation-by-distance model. Overall, based on RAPD markers, the level of nuclear genome variation in endemic *M. decussata* (Table) is comparable with that found in some other Cupressaceae species with fragmented or restricted ranges (e.g., Hwang et al., 2001; Allnutt et al., 2003; Hao et al., 2006).

Based on sequence data from noncoding cpDNA regions, a considerable level of haplotype diversity and a low level of nucleotide diversity have been revealed in *M. decussata*, which is similar to what has been found in some other woody species, including two species of Cupressaceae, *Cunninghamia konishii* and *C. lanceolata* (Hwang et al., 2003). A common haplotype for all populations has not been found, though three haplotypes of the central Sikhote Alin population are also present in southern and/or northern populations. The large number of unique, closely related haplotypes within each population may suggest that the distribution area of *M. decussata* was fragmented a long time ago by the extirpation of populations in the adjacent territory.

In contrast to nuclear DNA, there is no significant isolation-by-distance effect observed in the plastid genome in this species, but cpDNA data related to differentiation show a nonrandom geographical distribution of haplotypes. Differentiation in *M. decussata* appears to be associated with historical events and the complex mountain topography of its range. The results of nested clade analysis and coalescent simulation data provide evidence of species expansion (Artyukova et al., 2009). The presence of the same substitutions and shared haplotypes in populations from opposite ends of the range indicate that ancestral populations of this species might have formerly exhibited a contiguous distribution range. The highest gene diversity in both genomes and the presence of shared haplotypes in the population from the central Sikhote Alin, which coincides in position with the assumed site of this species' origin at the watershed of the ancient Ussuri and Partizanskaya River basins (Urusov, 1979), may indicate its ancient origin. The fossil data (see above) confirm the occurrence of *M. decussata* in the Oligocene/Pliocene plant associations of Sikhote Alin near the southern limits of the current species range.

Unique species life-history traits have ensured the survival and range expansion of *M. decussata*, while most Arcto-Tertiary species, including species of Cupressaceae, either shifted their ranges south (e.g., *Platycladus orientalis*) or vanished (e.g., putative common ancestor of *M. decussata* and *P. orientalis*). The long-term persistence of *M. decussata* in the territory that has been ice-free during glaciations as well as traits such as a long life span and pre-reproductive phase, the long-term survival of its seeds in soil seed banks and spreading by layering, seem to enable the retention of historically established levels of gene diversity in fragments of the ancestral populations of this species. Climatic and landscape changes at the Pleistocene–Holocene boundary caused the timberline to rise, and *M. decussata*, like most contemporary montane plants, retreated toward higher elevations. The enrichment of plant communities of the Sikhote Alin Mountains with cool temperate and boreal species also resulted in the shifting of *M. decussata* stands to ecotopes with severe climatic and soil conditions (e.g., steep, stony slopes and scree fields) and in the splitting of the large ancestral population, forming its disjunct present-day distribution.

The modern populations of *M. decussata* have the ability to survive the ongoing climate changes and global warming because of the physiological and ecological range of tolerance and life history traits of this species. Upward shifts of the timberline forcing *M. decussata* to migrate up to mountain summits with taluses (Urusov, 1988; Krestov & Verkholat, 2003) may lead to further contraction of its remnant populations on mountains at middle elevations. However, because this species can survive under harsh environmental conditions (e.g., poor soil, high solar radiation, extreme winds), which are unsuitable for other species, it could survive on the highest mountain peaks.

2.5 *Oxytropis chankaensis* Jurtzev

Oxytropis chankaensis Jurtz. (synonym *O. hailarensis* subsp. *chankaensis* (Jurtzev) Kitag.) is a perennial herb with a narrow habitat range that is restricted to the west shore of Khanka Lake (Kharkevich & Kachura, 1981; Pavlova, 1989; Yurtsev, 1964), which is the largest lake in Northeast Asia (Figure). *O. chankaensis* plants occur only in sandy habitats on a narrow strip along the Khanka Lake shoreline, forming separate populations numbering approximately 80 to 500 individuals (Kholina & Kholin, 2006). The genus *Oxytropis* DC is comprised of approximately 450 species occurring predominantly in the mountains of Asia. The high ecotopic diversity and the mosaic of conditions in mountain ecosystems lead to the occurrence of neighbouring populations of *Oxytropis* species with different ecological requirements and contribute to the enhanced speciation and interspecific hybridization in the genus. These processes explain the current controversy within the taxonomy of this genus (Malyshev, 2007, 2008).

A legume species originally found on the shoreline of Khanka Lake was first described as a distinct endemic species, *Oxytropis chankaensis*, based on the observation of definite morphological differences compared with its congeners (Yurtsev, 1964). In the International Legume Database (ILDIS), *O. chankaensis* is considered to be a subspecies of *O. hailarensis* Kitag., the species that occurs in China and Mongolia (Bisby et al., 2009). In the Flora of China (Zhu et al., 2010), both species are regarded as synonyms of *O. oxyphylla* (Pall) DC, a type species of section *Baicalia*, subgenus *Oxytropis* (Malyshev, 2007). However, *O. chankaensis* and *O. oxyphylla* are clearly distinguished by morphological features and by their ploidy levels: *O. oxyphylla* is diploid (2n = 16; Zhu et al., 2010), whereas *O. chankaensis* is tetraploid (2n = 32; Probatova et al., 2008a). Phylogenetic analyses of the ITS rDNA and three noncoding regions of chloroplast DNA strongly confirm that *O. chankaensis* and *O. oxyphylla* are distinct species (Artyukova et al., 2011a).

O. chankaensis is the only representative of section *Baicalia* in Primorye (Pavlova, 1989). It is an outcrossing species that is pollinated by bumblebees, like most *Oxytropis* species (Yurtsev & Zhukova, 1968), with pollen potentially being dispersed over long distances. *O. chankaensis* is characterized by a long flowering period (from the third week of May until mid-August) and a high number of flowers per plant (up to 400, an average of 40 inflorescences with 4–14 flowers per plant) as well as high pollen fertility (95.7 ± 1.4%), which contribute to successful pollination and fertilization. The plants of this species exhibit high fecundity, with fruits containing up to 20 seeds, and an individual plant produces approximately 4,000 seeds (Kholina et al., 2003). The first fruits mature by the end of May, and fruiting lasts until September. Mature spherical pods can be dispersed by wind and water over long distances beyond the limits of local populations, while some seeds from dehiscent pods are gravity-dispersed over only a short distance from the maternal plant to form the soil seed bank. During the vegetative season, seedlings emerge from the seeds of the first fruits, and some of the seeds begin to germinate in the following year, after a winter dormancy period.

Ontogenic features of *O. chankaensis* (Kholina & Kholin, 2006), such as its long life span, overlapping generations, the multiplicity of its development, its early transition to the generative state and the long period of the generative state, are the most important characteristics of this species for maintaining its population numbers and preserving genetic heterogeneity. The juvenile and immature stages are the most vulnerable stages of the species' life cycle.

The characteristics of the reproductive biology of *O. chankaensis*, such as the normal structure and function of the reproductive organs, the high fertility of pollen, the considerable duration of the flowering period and the long life of the flower, result in reliable pollination and high seed production. This species is characterized by a combination of different modes of dissemination, hardseededness and long-term maintenance of germination (over 10 years). Seed dispersal over short and long distances by wind and water promotes intraspecific genetic structure and the homogenization of populations through gene exchange. The longevity of the seeds leads to their accumulation in the seed bank in the soil, and these seeds replenish the gene pool of a population when they germinate. The high fecundity of *O. chankaensis*, together with mechanisms that support recombination (the predominance of cross-pollination) and the exchange of genes (via pollen and seeds), provides reliable renewal of this species *in vivo*.

An allozyme analysis showed that *O. chankaensis* is an autotetraploid that arose through the fusion of nonreduced gametes in the course of multiple crosses between genetically different plants (Kholina et al., 2004). The recurrent polyploidy events in the evolutionary history of this species are confirmed by the presence of several chloroplast DNA haplotypes in each population (Artyukova et al., 2011b). In addition, the levels of genetic diversity in *O. chankaensis* populations revealed using allozymes (Kholina et al., 2009) and RAPDs are high compared with the average values for rare endemic species (Table).

The most striking feature of the *O chankaensis* plastid genome, which is maternally inherited in Fabaceae, is the unexpectedly high cpDNA haplotype diversity (Table) for a species with an extremely narrow geographic range. At the same time, its nucleotide diversity is low. Most (≥90%) of the genetic variability of nuclear markers and of the chloroplast markers is distributed within populations. Unlike most angiosperms (Petit et al., 2005), the level of cpDNA subdivision in this species does not differ from the levels of differentiation of the nuclear genome based on RAPD markers (Table). In addition, correlation between geographic and genetic distances is absent for the plastid and nuclear genomes. The low population partitioning (cohesive genetic system) and lack of phylogeographic patterning may be attributed to both recent fragmentation of a once continuous population and extensive (past/modern) gene flow via pollen and seeds, which prevents the accumulation of genetic differences.

Despite the slight divergence observed, each population possesses a unique part of the species gene pool. The biological traits and reproductive features of this species as well as the existence of tetrasomic inheritance and recurrent tetraploidy events in its evolutionary history contribute significantly to the maintenance of genetic diversity.

Thus, the narrowly endemic species *O. chankaensis* exhibits adaptive mechanisms that enable it not only to successfully renew its populations in the coastal zone, which are exposed to frequent flooding and other adverse factors, but also to maintain the high level of recombination responsible for the survival of the species in a changing environment. Apparently, the rarity of this species is a result of its high habitat specificity; it lives only on the sandy shores of a large lake where there is intense insolation and high air humidity. Fluctuations in its population size as a result of lake-level oscillations can result in reduced numbers at some localities, but in these cases, the species' high productivity and reserves of genetic variability, which enable adaptive responses of species, help to restore the populations. The threat of total destruction of the species is raised by human-induced habitat destruction.

2.6 *Iris mandshurica* Maximowicz and *I. vorobievii* N.S. Pavlova

Iris L. is a Northern Hemisphere genus of flowering plants composed of approximately 280 valid species. As they are mostly open land plants, *Iris* species are adapted for living in a wide range of habitats from cold and montane regions to grassy slopes, steppe meadowlands, arid and marsh areas and riverbanks, though there are no truly sylvestral plants among this group. Irises are outcrossing species exhibiting flowers with specialized structures to insure cross-pollination; their fruits (dry capsule) contain numerous seeds that can disperse through a variety of mechanisms, such as barochory, autochory, anemochory, hydrochory, myrmecochory and zoochory. In addition, these perennial plants reproduce asexually through bulbs or rhizomes that form dense or loose colonies (tufts) and grow in size over 20 years. Many *Iris* species inhabiting Russia are at the border of their geographical ranges and occur in small, isolated populations. In the southern part of the Russian Far East, there are 11 *Iris* species, mainly belonging to the subgenus *Limniris*, section *Limniris* (Pavlova, 1987). Only two species, *Iris mandshurica* Maxim. and *I. vorobievii* N.S. Pavlova, are representatives of a small section of dwarf irises, *Psammiris*, of the subgenus *Iris*. Psammirises usually have yellow flowers with a yellow beard in the center of the outer three perianth segments ("falls") and seeds with a white appendage (loosely called aril) indicative of possible dispersal by ants.

Psammirises are mainly Asian species, and only *Iris arenaria*, which is considered a synonym of *I. humilis*, is widely distributed in southern Europe (Alexeeva, 2008). *I. mandshurica* and *I. vorobievii*, which are found in the south of Primorye, are very similar to *I. humilis*, but they differ from each other and from *I. humilis* in some morphological features (Pavlova, 1987; 2006; Alexeeva & Mironova, 2007; Bezdeleva et al., 2010) and in their chromosome numbers (Probatova et al., 2008b; Shu, 2000). The distinctiveness of these species has also been confirmed by the use of molecular DNA markers (Kozyrenko et al., 2009).

In Russia, small and disjunct populations of *I. mandshurica* occur rarely (Figure, arrows) on dry grassy slopes, stone hills and in the steppe meadow. *I. vorobievii* is found in the only locality in the extreme south of Primorye (Figure, arrow), where it grows on open grassy slopes of hills and in the meadows of fluvial terraces as low-density, isolated patches (Pavlova, 2006; Alekseeva, 2008). Outside Russia, *I. mandshurica* mainly occurs at altitudes of ca. 400–800 m in northeast China (Heilongjiang, Jilin and Liaoning Provinces) and in northern Korea (Figure; Shu, 2000), whereas there are no available data on the range of *I. vorobievii*, though the occurrence of the species in adjacent regions in China and Korea (Figure, strips) cannot be ruled out.

Like all *Iris* species, *I. mandshurica* and *I. vorobievii* reproduce sexually and asexually through rhizomes. The seeds of both species are gravity-dispersed to only a short distance from a maternal plant, though secondary dispersal by ants cannot be excluded entirely. *I. mandshurica* exhibits a thick, shortened rhizome growing almost horizontally and forming loose turf. The vertical, stout rhizome of *I. vorobievii* is very short (1 cm in length) and presents only a few lateral buds; the rhizome grows at one end while the old part of it dies, and the rhizome does not reach great lengths (Bezdeleva et al., 2010). These features indicate that the species is a short-lived perennial (Alexeeva, 2008), living no more than 5–7 years. In contrast to most *Iris* species, *I. vorobievii* is difficult to cultivate.

Both psammirises studied exhibit similar levels of nuclear and plastid DNA diversity (Table). Based on RAPD markers, the genetic diversity in populations of *I. mandshurica* and *I. vorobievii* corresponds with that in natural populations of the rare and endangered European steppe plant *Iris aphylla* ($P = 30.6\%$, $He = 0.097$; Wróblewska & Bzosko, 2006) but is lower than that in the widespread species *I. humilis* ($P = 48.1\%$, $He = 0.168$; Kozyrenko et al., 2009). Notably, the genetic diversity value in *I. humilis* is in accord with the average value for plants with mixed breeding systems ($He = 0.18$; Nybom, 2004), whereas in *I. mandshurica* and *I. vorobievii*, these values are significantly lower (Table). This may indicate the prevalence of vegetative propagation over propagation through seeds in the populations studied.

I. mandshurica and *I. vorobievii* show considerable levels of haplotype diversity, along with low levels of nucleotide diversity in the chloroplast genome, which is transmitted through seeds in *Iris* species (Cruzan et al., 1993). Low nucleotide diversity and high levels of cpDNA haplotype diversity have been found in some widespread (e.g., *I. humilis*, Kozyrenko et al., 2009) and endemic perennial herb species (e.g., *Aconitum gimnandrum*, Wang et al., 2009; and *Oxytropis chankaensis*, Artyukova et al., 2011b). However, for most angiosperms, including *Iris* species (e.g., Cornman & Arnold, 2007), populations are often fixed for single cpDNA haplotypes, and polymorphic populations possessing different haplotypes occur in potential contact zones of different maternal lines or at sites of long-term persistence. In contrast, the populations of *I. vorobievii* and *I. mandshurica* contain 13 and 6 haplotypes, respectively, which may result from historical gene flow, retention of ancestral polymorphisms that accumulated over a long period in more continuous ancient populations, or a putative origin from several founders. Based on cpDNA haplotypes, demographic event analyses show that populations of both species have undergone bottleneck events and expansion in the past (Kozyrenko et al., 2009).

Apparently, the main reason for the rarity of *I. mandshurica* and especially of *I. vorobievii* is the scarcity of suitable habitats for these psammirises in Primorye. As *I. vorobievii* and *I. mandshurica* represent components of steppe vegetation and grow in specific edaphic conditions, they are members of several steppe communities that are likely remnants of previously more widespread steppe vegetation (Krestov & Verkholat, 2003). The contractions of such relic communities result from climate and natural community changes (natural succession) and, in recent years, from anthropogenic habitat destruction. Their isolation from the main part of the species' ranges, limited seed dispersal and poor vegetative reproduction make these psammirises particularly vulnerable and may lead the species to extinction.

3. Conclusion

It is usually assumed that a certain level of genetic diversity is necessary for the long-term prosperity of a species. Indeed, different genotypes confer different levels of resistance to various environmental stresses, and consequently, the greater the diversity of the genotypes in a population, the more effective its ability to withstand unfavorable conditions will be. Therefore, the existence of a low level of genetic diversity is often considered to represent a crucial stage for the survival of a species or even a sign of its extinction. However, despite several attempts to determine the average levels of genetic variation in different categories of plants (e.g., Gitzendanner & Soltis, 2000; Nybom, 2004), it is still unknown what level of polymorphism should be considered critical for the existence of a certain species. Thus, some rare endemic species exhibit higher levels of genetic diversity than the average values found for this category of plants, such as *Oxytropis chankaensis* (Table). At the same time, very low levels, or even an absence of genetic variability has been found in other narrowly endemic plants (e.g., *Bensoniella oregona*, Soltis et al., 1992). It is impossible to predict the fate of a species on the basis of its genetic diversity alone. Here, we have attempted to address some species existing under extreme conditions in the context of a variety of environmental, biological, evolutionary and other influences.

All of the species described above are represented by small, fragmented, marginal populations. Each of these species is characterized by a particular level of genetic diversity, ranging from very low (*Panax ginseng*) to high (*Oxytropis chankaensis*). Despite these differences, all of these plants exhibit weak competitiveness, and none of them can be considered prosperous. These species are not dominant in their respective plant communities, and they inhabit specific (often quite narrow) ecological niches where their existence is maintained more or less successfully for a long period. Without the effects of human activity, these species could probably exist in this state indefinitely. What are the mechanisms that ensure the long-term existence of these species in small, isolated populations at the limit of their climatic and environmental tolerance?

For ginseng, longevity appears to be most important factor for its survival. Indeed, a lifetime of up to several hundred years is unusual for an herbaceous plant that is incapable of vegetative reproduction. Ginseng maintains the ability to produce seeds throughout its life span, and it appears that as a result of apomixes, its seed production is not particularly dependent on environmental conditions or the availability of pollinators. A large number of fully viable seeds and an extended period of embryo maturation allow revival of populations of this species, even from a small number of individuals over the course of many years.

Similar to ginseng, *Aristolochia manshuriensis* only exhibits seed-based reproduction. The life span of this species is not as long as that of other woody plants and is near the average for woody vines. This species has a very poor ability to reproduce vegetatively, and there are major limitations on its seed reproduction as a result of adaptation to pollination only by certain insects. Though adaptation to specific pollinators prevents inbreeding, pollination decreases if pollinators are absent, and the resulting fruit development is low. In this case, high seed production (a large number of seeds per one fertilization event) and adaptability to seed transfer by wind and water seems to be of primary importance for the survival of the species. Even a single successful fertilization every several years could guarantee population restoration if conditions are suitable for seed germination. Under unfavorable conditions, plantlets of this species can exist in a juvenile state for a long time, which allows a population to survive until the re-establishment of suitable conditions.

The ability for vegetative propagation is a pathway for the survival of some species (*Oplopanax elatus*, irises and *Microbiota decussata*). This pathway also allows for the rehabilitation of a species, even after a significant reduction in its population, and it effectively allows clonal colonies to reoccupy their habitats, where suitable conditions for the species are associated with an absence of competitors. The flexibility of the reproductive system of *O. elatus*, combined with its different modes of reproduction allows it to renew heterozygous genotypes by clonal growth and to contribute additional variability resources through sporadic seed reproduction. At the same time, some features of the biology of this species (the long life of a single clone, overlapping generations and the ability to cross-pollinate) also help to maintain a certain level of polymorphism.

For *I. mandshurica* and *I. vorobievii*, vegetative reproduction (even if poor) seems to be the only way to survive under unfavorable conditions. Rare reproduction through seeds allows for the maintenance of a certain level of genetic diversity in its populations. However, the spread of both species beyond the borders of their existing populations is unlikely as a result of the scarcity of suitable habitats. Poor vegetative reproduction and limited seed dispersal as well as isolation from the main part of the species range (*I. mandshurica*) or its occurrence in a single locality (*I. vorobievii*) make populations of both species vulnerable. In the case of *I. vorobievii*, this vulnerability may lead to rapid species extinction.

The pattern of genetic diversity and the structure of populations in the endemic species *M. decussata* are congruent with the leading edge model of colonization. It could be proposed that extant southern populations represent the putative species range center. Populations that expanded southward during dry and cool periods at the Oligocene/Miocene boundary have become completely extirpated following climate changes in the Quaternary. The distribution area of *M. decussata* became fragmented quite some time ago through the displacement of the species toward mountain peaks and the extinction of stands in the adjoining territories. The remnant *M. decussata* populations have the potential for survival under ongoing climate changes and global warming because of this species' physiological and ecological range of tolerance, reproduction by layering and other life history traits.

In the case of endemic or rare species, one must distinguish between relics left by the extinction of related populations and newly evolved taxa. All relics that survived the repeated periods of Pleistocene climate cooling have apparently experienced a severe genetic bottleneck, not only as a result of genetic drift associated with the reduction of populations, but also because of selection acting in a rapidly changing climate. Because of this selection, the fittest individuals have survived. Among the representatives of the ancient tropical floras (*P. ginseng, O. elatus, A. manshuriensis*) only a small number of genotypes are likely to have possessed such fitness. By acquiring a mechanism for survival in harsh environments, these species have lost much of their genetic diversity.

In the case of the narrowly endemic species *O. chankaensis*, the situation is different. This species possesses adaptive mechanisms that enable it not only to successfully renew its populations in the coastal zone, which are exposed to frequent flooding and other adverse factors, but also to maintain the high level of recombination responsible for the survival of the species in a changing environment. This is a relatively young species, and its biological characteristics may promote its prosperity and wide distribution; the only obstacle to this is its high habitat specificity; it only inhabits the sandy shore of a large lake where there is intense insolation and high air humidity. However, there are no such habitats nearby, and the small number of plants in some populations makes this species vulnerable.

The patterns observed for various rare species often do not fully correspond to the general idea of survival at the edge of their range. Plant species exhibit tremendous variation in life history traits that may help them survive in harsh or changeable environments. The cause of a plant's rarity depends on the effects of different historical, biological and genetic factors. In all cases, different compensatory mechanisms, such as increased longevity and fertility, the formation of soil seed banks and vegetative reproduction, are involved. The adaptations of these species are not always successful because the historically established balance between reproduction and dispersal can be disturbed. However, in the absence of destructive human activities, many rare species could exist for an indefinite time.

4. Acknowledgements

We are grateful to Dr. E.V. Sundukova and Dr. V.N. Makarkin for help with the manuscript preparation. This work was supported by the Grant "Molecular and Cell Biology" of the Russian Academy of Sciences (No 09-I-P22-03), by the Grant "Biological Diversity" of the Russian Academy of Sciences (No 09-I-P23-06), by the Grant "Biological Resources of Russia: Estimation and Fundamental Basis for Monitoring" (No 09-I-OBN-02), by the Grant No 11-III-B-06-094 of FEBRAS and by Grant "Leading Schools of Thought" from the President of Russian Federation.

5. References

Adams, C.A., Baskin, J.M. & Baskin, C.C. (2005). Trait Stasis Versus Adaptation in Disjunct Relict Species: Evolutionary Changes in Seed Dormancy–Breaking and

Germination Requirements in a Subclade of *Aristolochia* Subgenus *Siphisia* (Piperales). *Seed Sci. Res.*, Vol.15, No.2, (June 2005), pp. 161–173, ISSN 0960-2585

Alexeeva, N.B. (2008). Genus *Iris* L. (Iridaceae) in the Russia. *Turczaninowia*, Vol.11, No.2, (June 2008), pp. 5–68, ISSN 1560-7259

Alexeeva, N.B. & Mironova, L.N. (2007). Critical Notes on Some Species of the Genus *Iris* (Iridaceae) in Siberia and the Far East. *Bot. Zhurn.*, Vol.92, No. 6, (June 2007), pp. 916–925, ISSN 0006-8136

Allnutt, T.R., Newton, A.C., Premoli, A. & Lara, A. (2003). Genetic Variation in the Threatened South American Conifer *Pilgerodendron uviferum* (Cupressaceae), Detected Using RAPD Markers. *Biol. Conserv.*, Vol.114, No.2, (December 2003), pp. 245–253, ISSN 0006-3207

Artyukova, E.V., Gontcharov, A.A., Kozyrenko, M.M., Reunova, G.D. & Zhuravlev, Yu.N. (2005). Phylogenetic Relationships of the Far Eastern Araliaceae Inferred from ITS Sequences of Nuclear rDNA. *Russ. J. Genet.*, Vol.41, No.6, (June 2005), pp. 649–658, ISSN 1022-7954

Artyukova, E.V. & Kozyrenko, M.M. (2009). Genetic Diversity in the Endemic of Sikhote Alin, *Microbiota decussata* Kom. (Cupressaceae), Inferred from Nuclear and Chloroplast Genome Data, *Proceedings of the III International Conference "Mountain ecosystems and their components, Soil and Vegetation of Mountain Territories"*, pp. 119–125, ISBN 978-5-87317-581-9, Association of scientific publications of KMK, Moscow, Russia, August 24–29, 2009

Artyukova, E.V., Kozyrenko, M.M., Gorovoy, P.G. & Zhuravlev, Yu.N. (2009). Plastid DNA Variation in Highly Fragmented Populations of *Microbiota decussata* Kom. (Cupressaceae), an Endemic to Sikhote Alin Mountains. *Genetica*, Vol.137, No.2, (November 2009), pp. 201–212, ISSN 0016-6707

Artyukova, E.V., Kozyrenko, M.M., Kholina, A.B. & Aistova, E.V. (2011a). Genetic Relationships of *Oxytropis chankaensis* Jurtz. and *O. oxyphylla* (Pall.) DC. (Fabaceae) Inferred from Analyses of Nuclear and Chloroplast Genomes, *Proceedings of International symposium "Modern achievements in population, evolutionary, and ecological genetics"*, p. 10, ISBN 978-5-7442-1512-5, Vladivostok, Russia, June 19–24, 2011

Artyukova, E.V., Kozyrenko, M.M., Kholina, A.B. & Zhuravlev, Yu.N. (2011b). High Chloroplast Haplotype Diversity in the Endemic Legume *Oxytropis chankaensis* May Result from Independent Polyploidization Events. *Genetica*, Vol.139, No.2, (February 2011), pp. 221–232, ISSN 0016-6707

Bezdeleva, T.A., Mironova, L.N. & Dudkin, R.V. (2010). Features of Structural Organization of Some Species of the Genus *Iris*. *Bulletin of the Botanical Garden Institute FEB RAS*. No.5, pp. 21–25, ISSN 2222-5579

Bisby, F.A., Roskov, Y.R., Orrell, T.M., Nicolson, D., Paglinawan, L.E., Bailly, N., Kirk, P.M., Bourgoin, T. & Baillargeon, G. (Eds.). (2009). *Species 2000 & ITIS Catalogue of Life: 2009 Annual checklist*. Available from http://www.catalogueoflife.org/annual-checklist/2009/

Blokhina, N.I. (1988). Fossil Woods, In: *Cretaceous–Palaeogene of the Lesser Kuril Islands. New Data on Palaeontology and Geological History*, V.A. Krasilov, N.I. Blokhina, V.S. Markevitch & M.J. Serova (Eds.), 26–40, Far Eastern Branch of the Academy of Sciences of the USSR, Vladivostok, Russia

Bondarenko, O.V. (2006). Fossil Woods from the Pliocene of Southern Primorye. *PhD Thesis*, Institute of Biology and Soil Science, Far Eastern Branch of the Russian Academy of Sciences, Vladivostok, Russia

Bulgakov, V.P. & Zhuravlev, Yu.N. (1989). Generation of *Aristolochia manshuriensis* Kom. Callus Tissue Cultures. *Rastitel'nye Resursy*, Vol.25, No.2, (May 1989), pp. 266–270, ISSN 0033-9946

Cornman, R.S. & Arnold, M.L. (2007). Phylogeography of *Iris missouriensis* (Iridaceae) Based on Nuclear and Chloroplast Markers. *Mol. Ecol.*, Vol.16, No.21, (November 2007), pp. 4585–4598, ISSN 0962-1083

Cruzan, M.B., Arnold, M.L., Carney, S.E. & Wollenberg, K.R. (1993). cpDNA Inheritance in Interspecific Crosses and Evolutionary Inference in Louisiana Irises. *Am. J. Bot.*, Vol.80, No.3, (March 1993), pp. 344–350, ISSN 0002-9122

Ellstrand, N.C. & Roose, M.L. (1987). Patterns of Genotypic Diversity in Clonal Plant Species. *Am. J. Bot.*, Vol.74, No.1, (January 1987), pp. 123–131, ISSN 0002-9122

Figueroa-Esquivel, E.M., Puebla-Olivares, F., Eguiarte, L.E. & Núñez-Farfán J. (2010). Genetic Sructure of a Bird-Dispersed Tropical Tree (*Dendropanax arboreus*) in a Fragmented Landscape in Mexico. *Revista Mexicana de Biodiversidad*, Vol.81, No.3, (November 2010), pp. 789–800, ISSN 1870-3453

Gadek, P.A., Alpers, D.L., Heslewood, M.M. & Quinn, C.J. (2000). Relationships Within Cupressaceae Sensu Lato: a Combined Morphological and Molecular Approach. *Am. J. Bot.*, Vol.87, No.7, (July 2000), pp. 1044–1057, ISSN 0002-9122

Gitzendanner, M.A. & Soltis, P.S. (2000). Patterns of Genetic Variation in Rare and Widespread Plant Cogeners. *Am. J. Bot.*, Vol.87, No.6, (June 2000), pp. 783–792, ISSN 0002-9122

Grushwitsky, I.V. (1961). *Ginseng: the Aspects of Biology*, Nauka, Leningrad, Russia

Hampe, A. & Petit, R.J. (2005). Conserving Biodiversity under Climate Change: the Rear Edge Matters. *Ecol. Lett.*, Vol.8, No.5, (May 2005), pp. 461–467, ISSN 1461-0248

Hao, B., Li, W., Linchun, M., Li, Y., Rui, Z., Mingxia, T. & Weikai, B. (2006). A Study of Conservation Genetics in *Cupressus chengiana*, an Endangered Endemic of China, Using ISSR Markers. *Biochem. Genet.*, Vol.44, No.1-2, (February 2006), pp. 31–45, ISSN 0006-2928

Hewitt, G.M. (2004). Genetic Consequences of Climatic Oscillations in the Quaternary. *Philos. Trans. R. Soc. Lond. B. Biol. Sci.*, Vol. 359, No.1442, (February 2004), pp. 183–195, ISSN 1471-2970

Huang, S., Kelly, L.M. & Gilbert, M.G. (2003). *Aristolochia* Linnaeus., In: *Flora of China* (Ulmaceae through Basellaceae), 19.12.2003, Z.-Y. Wu, P.H. Raven & D.Y. Hong, (Eds.), Vol.5, , pp. 258–269, Available from http://www.efloras.org

Huh, M.K., Jung, S.D., Moon, H.K., Kim, S.-H. & Sung J.S. (2005). Comparison of Genetic Diversity and Population Structure of *Kalopanax pictus* (Araliaceae) and its Thornless Variant Using RAPD. *Korean Journal of Medicinal Crop Science*, Vol.13, No.2, (March 2005), pp. 69–74, ISSN 1225-9306

Hwang, S.-Y., Lin, H.W., Kuo, Y.S. & Lin, T.P. (2001). RAPD Variation in Relation to Population Differentiation of *Chamaecyparis formosensis* and *Chamaecyparis taiwanensis*. *Bot. Bull. Acad. Sin.*, Vol.42, No.3, (July 2001), pp. 173–179, ISSN 1817-406X

Hwang, S.-Y., Lin, T.P., Ma, C.-S. & Lin, C.-L. (2003). Postglacial Population Growth of *Cunninghamia konishii* (Cupressaceae) Inferred from Phylogeographical and Mismatch Analysis of Chloroplast DNA Variation. *Mol. Ecol.*, Vol.12, No.10, (October 2003), pp. 2689–2695, ISSN 0962-1083

Kim, C. & Choi, H.-K. (2003). Genetic Diversity and Relationship in Korean Ginseng (*Panax schinseng*) Based on RAPD Analysis. *Korean J. Genet.*, Vol.25, No.3, (September 2003), pp. 181-188, ISSN 0254–5934

Kim, S.H., Jang, Y.S., Han, J.G., Chung, H.G., Chung, H.G., Lee, S.W. & Cho, K.J. (2006). Genetic Variation and Population Structure of *Dendropanax morbifera* (Araliaceae) in Korea. *Silvae Genet.*, Vol.55, No.1, (January 2006), pp. 7–13, ISSN 0037-5349

Kitagawa, M. (1979). *Neo-Lineamenta Florae Manshuricae*, J. Cramer, ISBN 3-7682-1113-4, Hirchberg, Germany

Kharkevich, S.S. & Kachura, N.N. (1981). *Rare Plant Species of the Soviet Far East and Their Conservation*, Nauka, Moscow, Russia

Kholina, A.B., Markelova, O.V. & Kholin, S.K. (2003). Population Structure and Reproduction Biology of the Rare Endemic Species *Oxytropis chankaensis* Jurtz., *Proceedings of the XI Congress of the Russian Botanical Society "Botanical Researches in Asian Russia"*, pp. 369–370, ISBN 5-93957-063-1, Novosibirsk–Barnaul, Russia, August 18–22, 2003

Kholina, A.B., Koren, O.G. & Zhuravlev, Yu.N. (2004). High Polymorphism and Autotetraploid Origin of the Rare Endemic Species *Oxytropis chankaensis* Jurtz. (Fabaceae) Inferred from Allozyme Data. *Russ. J. Genet.*, Vol.40, No.4, (April 2004), pp. 393–400, ISSN 1022-7954

Kholina, A.B., Koren, O.G. & Zhuravlev, Yu.N. (2009). Genetic Structure and Differentiation of Populations of the Tetraploid Species *Oxytropis chankaensis* (Fabaceae). *Russ. J. Genet.*, Vol.45, No.1, (January 2009), pp. 70–80, ISSN 1022-7954

Kholina, A.B. & Kholin, S.K. (2006). Population Age Structure of Rare Plant *Oxytropis chankaensis*, *Proceedings of the 2 International Conference "Problems of Preservation of Wetlands of International Meaning: Khanka Lake"*, pp. 26–35, ISBN 5-91162-002-2, Spassk-Dalny, Russia, June 10-11, 2006

Kholina, A.B., Nakonechnaya, O.V., Koren, O.G., & Zhuravlev, Yu.N. (2010). Genetic Variation of *Oplopanax elatus* (Nakai) Nakai (Araliaceae). *Russ. J. Genet.*, Vol.46, No.5, (May 2010), pp. 555–561, ISSN 1022-7954

Koren, O.G., Potenko, V.V. & Zhuravlev, Yu.N. (2003). Inheritance and Variation of Allozymes in *Panax ginseng* C.A. Meyer (Araliaceae). *Int. J. Plant Sci.*, Vol.164, No.1, (January 2003), pp. 189–195, ISSN 1058-5893

Koren, O.G., Krylach, T.Yu., Zaytseva, Yu.A. & Zhuravlev, Yu.N. (1998). Floral Biology and Embryology of *Panax ginseng* C.A. Meyer, *Proceedings of the 1st European Ginseng Congress*, pp. 221-231, S. Imhof., Marburg, Germany, December 6–11, 1998

Koren, O.G., Nakonechnaya, O.V. & Zhuravlev, Yu.N. (2009). Genetic Structure of Natural Populations of the Relict Species *Aristolochia manshuriensis* (Aristolochiaceae) in Disturbed and Intact Habitats. *Russ. J. Genet.*, Vol. 45, No.6, pp. 678–684, ISSN 1022-7954

Kozyrenko, M.M., Artyukova, E.V. & Zhuravlev Yu.N. (2009). Independent Species Status of *Iris vorobievii* N.S. Pavlova, *Iris mandshurica* Maxim., and *Iris humilis* Georgi

(Iridaceae): Evidence from the Nuclear and Chloroplast Genomes. *Russ. J. Genet.,* Vol.45, No.11, (November 2009), pp. 1394–1402, ISSN 1022-7954

Krestov, P.V. & Verkholat, V.P. (2003). *Rare Plant Communities of Amur Region.* Institute of Biology and Soil Sciences, ISSN 5744213406, Vladivostok, Russia

Kurentsova, G.E. (1968). *Relic Plants of Primorye,* Nauka, Leningrad, Russia

Little, D.P., Schwarzbach, A.E., Adams, R.P. & Hsieh, C.-F. (2004). The Circumscription and Phylogenetic Relationships of *Callitropsis* and the Newly Described Genus *Xanthocyparis* (Cupressaceae). *Am. J. Bot.,* Vol.91, No.11, (November 2004), pp. 1872–1881, ISSN 0002-9122

Malyshev, L.I. (2007). Phenetics in the Section Verticillares of the Genus *Oxytropis* DC. (Fabaceae). *Bot. Zhurn.,* Vol.92, No.6, (June 2007), pp. 793–807, ISSN 0006-8136

Malyshev, L.I. (2008). Diversity of the genus *Oxytropis* in Asian Russia. *Turczaninowia,* Vol.11, No.3, (December 2008), pp. 5–141, ISSN 1560-7259

Melnikova, A.B. & Machinov, A.N. (2004). On the Record of *Microbiota decussata* (Cupressaceae) at Unusually Low Altitude. *Bot. Zhurn.,* Vol.89, No.9, (September 2004), pp. 1470–14720, ISSN 0006-8136

Nakonechnaya, O.V., Koren, O.G., Nesterova, S.V., Sidorenko, V.S., Kholina, A.B., Batygina, T.B. & Zhuravlev, Yu.N. (2005). Elements of Reproductive Biology of *Aristolochia manshuriensis* Kom. (Aristolochiaceae) in the Conditions of Introduction. *Rastitel'nye Resursy,* Vol.41, No.3, (August-September 2008), pp. 14–25, ISSN 0033-9946

Nakonechnaya, O.V., Koren, O.G. & Zhuravlev, Yu.N. (2007). Allozyme Variation of the Relict Plant *Aristolochia manshuriensis* Kom. (Aristolochiaseae). *Russ. J. Genet.,* Vol.43, No.2, (February 2007), pp. 156–164, ISSN 1022-7954

Nakonechnaya, O.V., Sidorenko, V.S., Koren, O.G., Nesterova, S.V. & Zhuravlev, Yu.N. (2008). Specific features of pollination in the Manchurian birthwort, *Aristolochia manshuriensis. Biol. Bull.* Vol.35, No.5, (October 2008), pp. 459–465, ISSN 1062-3590

Nechaev, V.A. & Nakonechnaya, O.V. (2009). Structure of Fruits and Seeds and Ways of Dissemination of Two Species of the Genus *Aristolochia* L. in Primorsky Krai. *Biol. Bull.,* Vol.36, No.4, (August 2009), pp. 393–396, ISSN 1062-3590

Nybom, H. (2004). Comparison of Different Nuclear DNA Markers for Estimating Intraspecific Genetic Diversity in Plants. *Mol. Ecol.,* Vol.13, No.5, (May 2004), pp. 1143–1155, ISSN 0962-1083

Pavlova, N.S. (1987). Family Iridaceae, In: *The vascular plants of the Soviet Far East,* Vol.2, S.S. Kharkevich, (Ed.), pp. 414–426, Nauka, Leningrad, Russia

Pavlova, N.S. (1989). Family Fabaceae. In: *The vascular plants of the Soviet Far East,* Vol.4, S.S. Kharkevich, (Ed.), pp. 191–339, Nauka, ISBN 5-02-026577-2, Leningrad, Russia

Pavlova, N.S. (2006). Family Iridaceae. *Iris* L., In: *Flora of the Russian Far East, Additions and Modifications to "Vascular Plants of the Soviet Far East",* A.E. Kozhevnikov, N.S. Probatova, (Eds.), pp. 277–279, Dal'nauka, ISBN 5-8044-0534-9, Vladivostok, Russia

Pavlyutkin, B.I., Petrenko, T.I. & Chekryzhov, I.Y. (2005). The Problems of Stratigraphy of the Tertiary Deposits of the Pavlovka Coal-field (Primorye). *Tikhookeanskaya Geologiya,* Vol.24, No.6, (November 2005), pp. 59–76, ISSN 0207-4028

Petit, R.J., Duminil, J., Fineschi, S., Hampe, A., Salvini, D. & Vendramin, G.G. (2005). Comparative Organization of Chloroplast, Mitochondrial and Nuclear Diversity in Plant Populations. *Mol. Ecol.,* Vol.14, No.3, (March 2005), pp. 689–701, ISSN 0962-1083

Probatova, N.S., Seledets, V.P. & Rudyka, E.G. (2008a). *Oxytropis chankaensis*. In: *Taxon* Marhold K. IAPT/IOPB chromosome data 5. Vol.57, No.2, (May 2008), pp. 560, E20, ISSN 0040-0262

Probatova, N.S., Rudyka, E.G., Seledets, V.P. & Nechaev, V.A. (2008b). *Iris vorobievii* In: *Taxon* Marhold K. IAPT/IOPB chromosome data 6. Vol.57, No.4, (November 2008), pp. 1270, E8, ISSN 0040-0262

Red Data Book, Primorsky Krai: Plants. (2008). AVK Apelsin, ISBN 978-5-98137-017-5, Vladivostok, Russia

Red Data Book of the Russian Federation: Plants. (1988). Rosagropromizdat, ISBN 5-260-00254-7, Moscow, Russia

Reunova, G.D., Kats, I.L., Muzarok, T.I. & Zhuravlev, Yu.N. (2010a). Polymorphism of RAPD, ISSR and AFLP Markers of the *Panax ginseng* C. A. Meyer (Araliaceae) Genome. *Russ. J. Genet.*, Vol.46, No.8, (August 2010), pp. 938–947, ISSN 1022-7954

Reunova, G.D., Kats, I.L., & Zhuravlev, Yu.N. (2010b). Genetic Variation of *Oplopanax elatus* (Araliaceae) Populations Estimated Using DNA Molecular Markers. *Doklady Biological Sciences*. Vol.433, No.1, (August 2010) pp. 252–256, ISSN 0012-4966

Sakaguchi, S., Sakurai, S., Yamasaki, M. & Isagi, Y. (2010). How Did the Exposed Seafloor Function in Postglacial Northward Range Expansion of *Kalopanax septemlobus*? Evidence from Ecological Niche Modeling. *Ecol. Res.*, Vol.25, No.6, (July 2010), pp. 1183–1195, ISSN 0912-3814

Schreter, A.I (1975). Medicinal Flora of the Soviet Far East. Medicine, Moscow, Russia

Shlotgauer, S.D. (2011). Characteristics of the Alpine vegetation of the Sikhote-Alin Range Illustrated by Ko Mountain (Khabarovsk Krai). *Contemp. Probl. Ecol.*, Vol.4, No.2, (April 2011), pp. 159–163, ISSN 1995-4255

Shu, Y.W. (2000). *Iris* Linnaeus. In: *Flora of China (Flagellariaceae through Marantaceae)*, n.d, Z.-Y. Wu, P.H. Raven & D.Y. Hong, (Eds.), Vol.24, pp. 297–313, Available from http://www.efloras.org

Shulgina, V.V. (1955). Woody Vines and Their Culture in Leningrad, In: *Introduction of Plants and Green Building*, S.Ya. Sokolov, (Ed.), pp. 157–194, Publishing House of Academy of Sciences, Moscow-Leningrad, Russia

Soltis, P.S., Soltis, D.E. & Tucker, T.L. (1992). Allozyme Variability Is Absent in the Narrow Endemic *Bensoniella oregona* (Saxifragaceae). *Conserv. Biol.*, Vol.6, No.1, (March 1992), pp. 131–134, ISSN 0888-8892

Urusov, V.M. (1979). Ecological and biological peculiarities of *Microbiota decassata* Kom. (Cupressaceae). *Bot. Zhurn.*, Vol.64, No.3, (March 2004), pp. 362–376, ISSN 0006-8136

Urusov, V.M. (1988). *Genesis of Vegetation and Conservancy in the Far East*. Far Eastern Branch of the Academy of Sciences of the USSR, Vladivostok, Russia

Vucetich, J.A. & Waite, T.A. (2003). Spatial Patterns of Demography and Genetic Processes Across the Species' range: Null hypotheses for landscape conservation genetics. *Conserv. Genetics*, Vol.4, No.5, (September 2003), pp. 639–645, ISSN 1572-9737

Wang, L., Abbott, R.J., Zheng, W., Chen, P., Wang, Y. & Liu, J. (2009). History and Evolution of Alpine Plants Endemic to the Qinghai-Tibetan Plateu: *Aconitum gimnandrum* (Ranunculaceae). *Mol. Ecol.*, Vol.18, No.4, (February 2009), pp. 709–721, ISSN 0962-1083

Watkinson, A.R. & Powell, J.C. (1993). Seedling Recruitment and the Maintenance of Clonal Diversity in Plant Population – a Computer Simulation of *Ranunculus repens*. *J. Ecol.*, Vol.81, No.4, (December 1993), pp. 707–717, ISSN 00220477

Wolf, A.T., Howe, R.W. & Hamrick, J.L. (2000). Genetic Diversity and Population Structure of the Serpentine Endemic *Calystegia collina* (Convolvulaceae) in Northern California. *Am. J. Bot.*, Vol.87, No.8, (August 2000), pp. 1138–1146, ISSN 0002-9122

Wróblewska, A. & Bzosko, E. (2006). The Genetic Structure of the Steppe Plant *Iris aphylla* L. at the Northern Limit of its Geographical Range. *Bot. J. Linn. Soc.*, Vol.152, No.2, (October 2006), pp. 245–255, ISSN 0024-4074

Xiang, Q.B. & Lowry, P.P. (2007). Araliaceae, In: *Flora of China*, 1.07.2007, Z.-Y. Wu, P.H. Raven, D.Y. Hong, (Eds.), Vol.13, pp. 435–491, Available from http://www.efloras.org

Xie, G.W., Wang, D.L., Yuan, Y.M. & Ge X.-J. (2005). Population Genetic Structure of *Monimopetalum chinense* (Celastraceae), an Endangered Endemic Species of Eastern China. *Ann. Bot.-London.*, Vol.95, No.5, (April 2005), pp. 773–777, ISSN 0305-7364

Yurtsev, B.A. (1964). Conspectus of the System of Section *Baicalia* Bge., Genus *Oxytropis* DC. In: *News in the systematics of higher plants*. I.A. Linchevskii, (Ed.), pp. 191–218, Nauka, Moscow-Leningrad, Russia

Yurtsev, B.A. & Zhukova, P.G. (1968). Polyploid Series and the Taxonomy (on the Basis of the Analysis of Some Groups of Arctic Legumes). *Bot. Zhurn.*, Vol.53, No.11, (November 1968), pp. 1531–1542, ISSN 0006-8136

Zhu, X.Y., Welsh, S.L. & Ohashi, H. (2010). Oxytropis, In: *Flora of China (Fabaceae)*, 16.02.2010, Z.-Y. Wu, P.H. Raven & D.Y. Hong, (Eds.), Vol.10, pp. 453–500, Available from http://www.efloras.org

Zhuravlev, Yu.N. & Kolyada, A.S. (1996). *Araliaceae: Ginseng and Others*, Dalnauka Press, ISBN 5-7442-0867-4, Vladivostok, Russia

Zhuravlev, Yu.N., Koren, O.G., Reunova, G.D., Muzarok, T.I., Gorpenchenko, T.Yu., Kats, I.L. & Khrolenko, Yu.A. (2008). *Panax ginseng* Natural Populations: Their Past, Current State and Perspectives. *Acta Pharmacol. Sin.*, Vol.29, No.9, (September 2008), pp. 1127-1136, ISSN 1671–4083

Zhuravlev, Yu.N., Reunova, G.D., Kats, I.L., Muzarok, T.I. & Bondar, A.A. (2010). Molecular Variation of Wild *Panax ginseng* C.A. Meyer (Araliaceae) by AFLP Markers. *Chin. Med.*, Vol.5, No.21, (June 2010), ISSN1749-8546 (Electronic), Available from http://www.cmjournal.org/

7

Founder Placement and Gene Dispersal Affect Population Growth and Genetic Diversity in Restoration Plantings of American Chestnut

Yamini Kashimshetty[1], Melanie Simkins[2],
Stephan Pelikan[3] and Steven H. Rogstad[1]
[1]Department of Biological Sciences
[2]Department of Environmental Studies
[3]Department of Mathematical Sciences
University of Cincinnati, OH
USA

1. Introduction

The American chestnut, *Castanea dentata* [Marsh.] Borkh (Fagaceae), was an abundant canopy tree inhabiting the mixed mesophytic forests of eastern North America. The species was struck by a fungal pathogen (*Cryphonectria parasitica* [Murrill] Barr) introduced from East Asia in the late 1800's on imported Asian chestnut material with the consequence that billions of trees have been destroyed (Barakat *et al.*, 2009; Pierson *et al.*, 2007; Elliot & Swank 2008; Jacobs 2007; Stilwell *et al.*,2003; Paillet, 2002; Huang *et al.*, 1998; Russel, 1987). The near elimination of this once important species has had widespread effects on the ecological functioning of eastern North American forests, and has also had a severe impact on economic forest extraction practices (e.g., strong workable lumber; chestnuts as food and forage). *Castanea dentata* has escaped complete elimination from its native range by persisting as occasional sprouts from the root collar of trees damaged by the blight; these sprouts rarely reach full sexual maturity (Jacobs, 2007; Stilwell *et al.*,2003; Paillet, 2002). Efforts are underway to restore this former keystone species and prevent extinction, reestablishing its ecological and economic roles in its natural habitat. Breeding of blight-resistant strains is being attempted (e.g., by the American Chestnut Foundation, TACF) through a series of initial hybridizations between *C. dentata* and the Chinese chestnut (*C. mollisima* Blume), followed by a series of backcrosses with the American chestnut always selecting for blight-resistance, to develop strains that are predominantly American chestnut in genotypic constitution but which retain Chinese blight-resistance genes. For example, some of the most recent blight-resistant strains ready for re-introduction are calculated to be genetically 94% American chestnut and 6% Chinese chestnut (Jacbos, 2007; Diskin *et al.*, 2006).

The conservation of endangered plant species often involves restoring these species back to their natural habitats and/or *ex situ* rescue plantings (Merritt & Dixon, 2011), and the American chestnut is no exception. Considerable resources are being expended in generating blight-resistant strains of *C. dentata*, thus necessitating optimization of restoration programs which are usually labor-intensive and expensive. Costs of such restoration efforts include propagule generation and collection, storage, treatment, site preparation, planting, protecting, provisioning, travelling to and from the introduction site, monitoring, and future manipulations of individuals (Rogstad & Pelikan, 2011).

One approach to optimizing restoration of the American chestnut is to use computer programs to model the population growth and genetic effects of restoring plant populations in different ways. While restoration programs have been widely undertaken for a number of plant species, we lack the tools to analyze how factors like founder number and geometry of placement within a restoration habitat interact with varying founder or species life history characteristics, and whether these factors impact resultant population growth rates and preservation of genetic diversity. While ecological edge effects have been documented as being evident at the borders and edges of restoration preserves due to altered micro-environments in terms of wind-speeds, light availability and organismal composition among other factors (Primack, 2010), less is known about whether placement of founders at varying distances from the preserve edge impacts the population dynamics and genetic diversity measures of establishing populations. Further, can differing founder placement patterns interact with other life-history characteristics such as pollen and offspring dispersal distances to bring about different population growth rates and genetic diversity levels? Exploring these potential demographic and genetic edge effects in the field with C. dentata would not be feasible due to the costs associated with carrying out such experiments on a wide-scale to ensure statistical reliability of data, especially since blight-resistant individuals are expensive and time-consuming to produce, and thus must be used judiciously. In this case modeling virtual populations through computer simulations represents a more tractable alternative to field experiments, potentially providing valuable insight for restoration managers on how best to re-introduce the American chestnut into preserves.

In this study, the computer program NEWGARDEN (Rogstad & Pelikan, 2011) was used to model blight-resistant American chestnut population restoration in a virtual preserve to explore the population growth and genetic effects of placing founders at different distances from preserve borders under differing patterns of gene dispersal. Our null hypothesis is: Varying American chestnut founder placement at various diagonal distances from the preserve border, while altering offspring and pollen dispersal distances, will have no effect on population growth rates or retention of founding genetic diversity. We used comparative trials to examine the degree to which some patterns of introduction might be preferable over others.

2. Methods

We used NEWGARDEN to model the population growth and genetic diversity of newly establishing chestnut populations. This program simulates natural population development based on a set of user-specified initial input conditions (Rogstad & Pelikan, 2011). One set of input conditions constitutes a "trial" and input between trials may vary as to founder number and geometric patterning, preserve characteristics, and with regard to various life-history characteristics (see below). For each age (bout of mating with mates chosen randomly as conditioned by the input) of the developing population, NEWGARDEN provides output statistics (for each new cohort and for the entire population; only the latter are given here) concerning the population size, the number of founding alleles retained, heterozygosity (observed and expected), and F_{it} (a measure reflecting deviation from Hardy-Weinberg expectations due to subdivision and/or inbreeding; hereafter referred to as F

Founder Placement and Gene Dispersal Affect
Population Growth and Genetic Diversity in Restoration Plantings of American Chestnut

133

value). For each set of input trial conditions, the user can specify the number of replicate runs of those conditions to be used to calculate a mean and standard deviation for each of the listed output statistics. The user can thus compare whether means of statistics differ for populations of contrasting trials that differ in one or more input conditions.

2.1 Input parameters held constant

The input parameters common to all experimental runs are described following their order in the NEWGARDEN input file. These constant input parameters are based on population development data obtained from the only known chestnut restoration stand which is naturally developing at West Salem, Wisconsin (for details, see Pierson *et al.*, 2007; Rogstad & Pelikan, 2011).

2.1.1 The preserve grid system

Populations for all trials develop on a Cartesian grid system defined by the user. Individuals can only establish on grid points, the spacing of which represents the average minimum distance that can exist between reproductive trees. In our simulations, the grid system represents a restoration preserve (a 5 km x 5 km square), comprising a 1000 x 1000 grid point system with 5 m between grid points. This simple model represents an approximate average minimum distance spacing needed between two reproductive trees based on a mature chestnut community at the carrying capacity.

2.1.2 Loci and allelic variation

The ideal, non-inbred source population for the founders was set to have 30 loci per founder, each with 100 alleles of equal frequency (frequency = 0.01 per allele). NEWGARDEN randomly draws two alleles for each locus from the source population when creating a founder's genotype. In each new generation, every individual was censused for its allelic status at each of the 30 loci to generate the genetic diversity output statistics.

2.1.3 Mating system

All individuals are bisexual with 100% outcrossing.

2.1.4 Offspring production

The age-specific reproduction rate (r) specified in all input files (based on population development at West Salem) was held constant as shown in Table 1. In the intermediate years not defined, the rate of offspring production is linearly interpolated between the bounding age values. An established offspring does not reach reproductive age until its eighth year (Table 1). Offspring production is distributed across eligible reproducing individuals according to a Poisson distribution. As in natural settings, reproductive individuals may create multiple offspring, but all of these may not establish into saplings. Offspring counted in a generation's reproduction rate are potential recruits for establishment and growth at a grid point. The number of such potential recruits does not indicate the exact size of the newly established cohort for any particular age in one bout of mating. See mortality for ways in which potential offspring "die" or fail to establish in NEWGARDEN.

2.1.5 Age-specific pollen rate

For established individuals, the relative rate of serving as an eligible pollen donor that contributes pollen to a given mating is conditioned by its age according to Table 1. Pollen production is proportional relative to the highest value in the input. For the ages that are not specified by the input values, the pollen rate is linearly interpolated. Once NEWGARDEN selects an age-class and distance class (see below) for the pollen donor, one donor is chosen at random for a particular mating from the pool of eligible potential donors.

2.1.6 Mortality

Individuals can "die" in a number of different ways during a trial run. Death means that the individual is removed from the grid system and from further population processes and analyses. If the offspring is distributed outside of the defined preserve grid system, it automatically dies. An offspring will die in the event that it lands on a grid point that is already occupied by an individual that will survive to the next generation. When an ovule isn't pollinated since an eligible pollen donor is not within the range of the ovule producer, this counts as a reproductive event followed by immediate death of the potential offspring. If two or more offspring land at the same grid point, one is randomly chosen as the survivor with others deleted. Death of individuals is also age-dependent as specifiable by the user. For all trials here, age-specific probabilities of dying are given in Table 1. Ages without a specified risk of death are linearly interpolated between bounding values. In our trials, founders reach age 114, so there is a probability that several founders are still alive by the end of the trial run.

Age	Relative rate of reproduction	Relative rate of pollen provisioning	Percent mortality
0	0	0	0.3
1	0	0	0.2
4	0	0	0.1
5	0	0	
6	0	0.2	
7	0		0.05
8	0.02		
10	0.04	0.5	0.02
12	0.08		
17	0.2		
25		0.95	
70	1.3		
75	1.3	1.2	
115	1.3	1.16	0.02

Table 1. Input Parameters Held Constant. User specified age-specific input conditions for reproductive rate, pollen provisioning, and mortality used in all trials. Values not given were interpolated between the bounding values. See text for more details.

Founder Placement and Gene Dispersal Affect
Population Growth and Genetic Diversity in Restoration Plantings of American Chestnut

135

2.1.7 Initial population

Aging of the population begins with the founders and it will continue to age until the specified number of generations is completed. 169 founders, this being approximately the minimum number of founders needed to capture > 95% of the source population alleles (Lawrence et al., 1995; Chakraborty, 1993), of initial age 13 were used with founders located every other grid point in a 13 x 13 individual square (26 x 26 grid points). This simulates a 10 m distance between founders in a planting square 130 m on each side. This square of founders remained constant across all trials, although it was moved to different distances from the preserve borders.

2.1.8 Generations

Beginning with the founders, 100 bouts of reproduction (101 generations or ages including the founding generation) were conducted for each set of trial conditions. Output results in Figures 1 to 3 are reported for the entire population at each age. For the pollen and offspring dispersal distance comparison trials (Figures 4 to 6) only the results at the terminal age 100 are given.

2.1.9 Replicate runs

The output values reported are mean values calculated from 40 replicate runs for each set of trial input conditions. When trials are said to differ, we mean that reported generation mean values differ significantly (p value $\leq$ 0.05) unless otherwise stated. This convention is used rather than reporting each standard deviation and t-test results for all trial comparisons.

2.2 Variable input parameters

2.2.1 Region and founder placement

In the first series of comparative trials, the square of 169 founders was placed at increasing diagonal distances from the preserve border beginning with the founders at the lower left corner (trial a). Seed and pollen dispersal distances were held constant at 25% to or from each distance frame (see below) in this initial series. In a subsequent series of trials, the constant-sized founder square was likewise moved diagonally toward the center of the preserve while offspring and pollen dispersal distances differed in various combinations (Table 2).

2.2.2 Offspring dispersal distribution

Offspring dispersal on the grid system is based on a nested series of limiting frames called "distance frames". When an offspring is dispersed to a particular distance frame based on given probabilities, the lower and upper values of each frame define the limits of dispersal. For a given dispersal event, one point within the selected distance frame is chosen at random. The limits of the four frames used in our trials and their varying respective dispersal probabilities across trials are described in Table 2. In the "Basic Conditions" trials, offspring establishment was evenly divided to the four distance frames with 25% probability of dispersal to each frame. Modifications to the "basic" trial conditions to create alternate comparative trials are detailed in Table 2. We use the summary phrase "Offspring less distant" to indicate restricted offspring dispersal (compared to the "Basic Conditions" trials) within 65 m of the parent (13 grid points), whereas in "Offspring least distant" trials, the majority of offspring dispersal was within 30 m (6 grid points) of the parent plant. The summary phrase "Offspring more distant" is used for the trials in which the greatest percentage of offspring were dispersed between distances 110 m to 1500m (22 to 300 grid points) from the parent plant (Nathan *et al.*, 2008; Jansen *et al.*, 2008).

2.2.3 Pollen transport distances

The "Basic Conditions" and "Offspring less/least/more distant" trials had equal probability of pollen dispersal from within each of the four distance frames (Table 2). Trial E has 90% of pollen dispersal limited to from within a 30 m frame of the producing individual. Trial F has 90% of pollen dispersed from 110 m to 1500 m (22 to 300 grid points) from the parent plant.

2.3 Output

Based on population characteristics developing from the initial input specifications, NEWGARDEN provides means for four statistical measures reported here: population size, mean number of founding alleles retained, observed heterozygosity, and F_{it} (or F value) calculated as:

$$F_{it} = 1 - H_{ob} / H_{ex} \tag{1}$$

where H_{ob} equals observed heterozygosity (based on actual counts of heterozygous loci across all loci for all the individuals in the population) and H_{ex} equals the Hardy-Weinberg expected heterozygosity based on the allele frequencies across all loci in the entire population. In general, F_{it} increases as inbreeding and/or subdivision increase in the population and $F_{it} = 0$ in the absence of inbreeding. For the first series of trials (Figures 1 through 3) mean output values are reported for the total number of individuals in the population at the end of each age. For the second series of trials (Figures 4 through 6), values are given only for the total population after 100 bouts of mating.

3. Results

Under the "basic" conditions of offspring and pollen dispersal (25% to or from each frame respectively; see Table 2), mean population size increased with increasing diagonal distance of the standard square of 169 founders from the preserve corner (Figure 1). At age 100, founders situated at the corner (trial a) had the lowest mean population size (2,018 individuals), while the equally highest mean population sizes (approximately 7,000) were attained by populations with founders inset by 300, 400 and 500 grid points from the corner (trials d, e, and f, respectively). Even an inset distance of 100 grid points (ca. 500 m; trial b) increased the mean population size by 148%, compared to founders placed at the corner. The percent gain in population increase per unit inset distance declines as the inset distances increases beyond 100 grid points, only increasing by approximately 21% when inset distances increases from 100 to 200 grid points, and approximately 13% when inset distances increase from 200 to 300 grid points. Beyond 300 grid points the rise in mean population size was not statistically different.

Founder Placement and Gene Dispersal Affect
Population Growth and Genetic Diversity in Restoration Plantings of American Chestnut

137

Trial Description	Trial	Frame dimensions (in grid points)							
		0-5		6-12		13-22		22-301	
		%Pollen	% Offspring	%Pollen	% Offspring	%Pollen	% Offspring	%Pollen	% Offspring
		from	to	from	to	from	to	from	to
Basic conditions	A	25	25	25	25	25	25	25	25
Offspring less distant	B	25	60	25	30	25	4	25	6
Offspring least distant	C	25	90	25	5	25	3	25	2
Offspring more distant	D	25	2	25	3	25	15	25	80
Pollen less distant, Offspring more distant	E	90	2	5	3	3	15	2	80
Pollen more distant, Offspring least distant	F	2	90	3	5	15	3	80	2

Table 2. Percent offspring and pollen dispersed to or from each distance frame for NEWGARDEN trials described further in the text. Trials were otherwise identical.

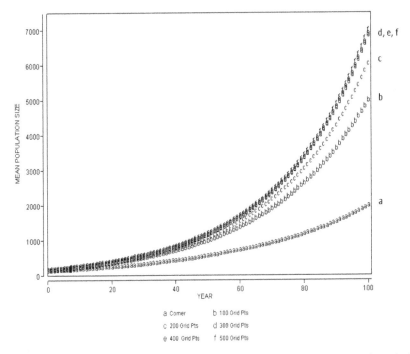

Fig. 1. Mean population sizes for each age of American chestnut populations founded at differing diagonal inset distances. These comparative trials span 101 generations at the basic offspring and pollen dispersal conditions (25% to and from each frame respectively). See text for more details.

Figure 2 shows the mean number of founding alleles retained across 101 generations for the trials depicted in Figure 1. In the source population, 3000 alleles each at frequency of 0.01, were available for the founding population. Drawing 169 founders, approximately 2900 alleles were present at founding (97% of the source population alleles; Figure 2). For example, imagine a restoration project where 1,690 trees are planted with 90% attrition prior to reproduction (e.g., Primack & Miao, 1992). On average such a population would have approximately 97% of the alleles in the original source population. NEWGARDEN can be used to estimate founding population sizes needed given the effects of attrition, to provide target numbers of individuals for establishing new populations or supplemention as needed.

For all populations, the number of alleles retained declined through generations, with populations founded at greater inset distances retaining higher numbers of alleles (Figure 2). The 169 founders were all aged 13 at the beginning of the trials, and given the mortality rate of 2% per generation (Table 1), approximately 22 founders should remain in populations after 100 bouts of mating. This suggests that a significant proportion of the total founding alleles remaining in the population at age 100 are carried by the descendents of founders that have died.

After 100 bouts of mating, there was approximately a 6% decline in number of founding alleles retained for populations inset at least 100 grid points or more (trials b, c, d, e and f), compared to the 9.6% decline seen for populations situated at the corner (trial a). At year 100, placing the founders 100 or more grid units inward from the corner produced a 3.8% (at the least; trials b through e) increase in alleles retained compared to placing the founders at the corner (trial a). The differences between various inset distances from 100 to 500 grid points (trials b through f) were not statistically different (although notice the trend). Variations of inset distances under basic dispersal conditions did not have significant or biologically meaningful effects on observed heterozygosity levels (ranging from 0.97 to 0.99) across generations, at basic dispersal conditions (data not shown).

There was little population subdivision and inbreeding observed across generations under basic dispersal conditions as founders were placed at greater inset distances (Figure 3). At year 100, F values increased slightly across generations, with the highest F values (almost 0.02) being reached by founders inset the furthest (by 500 grid points; trial f). F values tended to be lower for decreasing inset distances (e.g., trials a and b) at year 100. Although all trials are increasing in F, only by increasing the number of generations could it be determined the degree to which F values might become important (e.g., approach 0.05) in future generations.

Next, we ran a series of comparative trials varying not only in the inset distance of the square of 169 founders, but also altering the offspring and pollen dispersal conditions relative to the basic conditions (25% to or from each frame respectively) used in the trials just reviewed in Figures 1 through 3. We conducted this next trial series to investigate how differing types of gene dispersal might interact with founder distance from a border to affect population growth and genetic diversity.

Figure 4 shows population growth results (at population age 100 only) from these trials when both inset distances and dispersal conditions were varied. Trends observed here included increasing mean population sizes as inset distances increased from the corner to

Founder Placement and Gene Dispersal Affect
Population Growth and Genetic Diversity in Restoration Plantings of American Chestnut

139

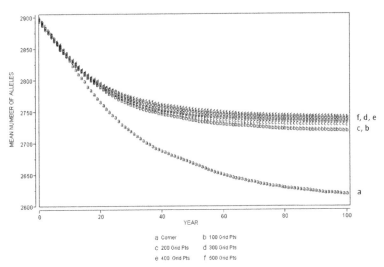

Fig. 2. Mean number of founding alleles retained across 101 generations for trial populations differing only in the distance to which founders were inset from a preserve border (population sizes shown in Figure 1). For all of these trials, the "basic" conditions of offspring and pollen dispersal distances were used (25% to or from each frame respectively). See text for more details.

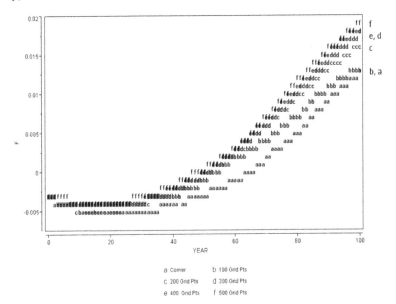

Fig. 3. F values across 101 generations for trials differing only in founder inset distance. These data are from the same trials depicted in Figures 1 and 2. "Basic" offspring and pollen dispersal conditions were used in each trial (25% to or from each frame respectively). See text for more details.

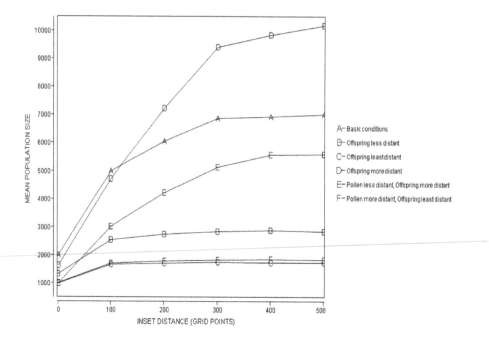

Fig. 4. Mean population size differences for comparative trial populations differing as to both founder inset distances (x-axis) and with greater or lower offspring and/or pollen dispersal distances (indicated by different letters A through F) relative to the "basic" dispersal conditions (see Table 2). Data points depict population size for each trial at age 100 only. In the trial condition summaries to the right of the graph, if a dispersule type (i.e., offspring versus pollen) is not mentioned, that dispersule disperses according to the basic conditions.

500 grid points for population D, up to 400 grid points (E), up to 300 grid points (A), and up to 100 grid points (B, C, and F). For populations B, C and F, increasing founder inset distance to 100 grid points from the corner raised mean population size significantly. However beyond 100 grid points, population sizes remained more or less constant across further increase in inset distances. If actual dispersal patterns match those in trials B, C, or F, a restoration manager will not gain higher rates of population growth by planting founders beyond 100 m into the preserve. Relative to the "basic" dispersal conditions trials just discussed in Figures 1 through 3 (trials marked A in Figure 4), altering offspring and pollen dispersal distances caused various but pronounced differences in rates of population growth. Trial D, with only the offspring dispersing to greater distances, was the only trial that exceeded trial A ("basic" conditions) at founder inset distances greater than 100 grid points. Trial D showed the greatest overall population growth compared to all trials at those inset distances.

For each set of trial conditions (A through F) the greatest allele loss occurred when founders were placed at the preserve corner (Figure 5). More alleles were retained under otherwise

Founder Placement and Gene Dispersal Affect
Population Growth and Genetic Diversity in Restoration Plantings of American Chestnut

141

constant trial conditions when founders were inset only 100 grid points although even further inset distances caused no further major increases in allele retention for trial conditions A, B, C, and F, while inset distances of 300 grid points were needed to maximize allele retention for trial conditions D and E. Considering all of these trial conditions, the greatest difference in the mean number of founding alleles retained is between trial E at the corner and trial D at 400 grid points (a 10.9% difference).

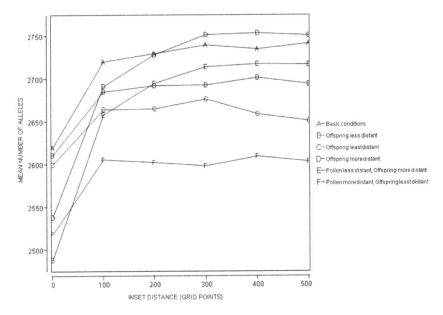

Fig. 5. Change in mean number of founding alleles retained in NEWGARDEN populations after 100 bouts of mating for trials depicted in Figure 4. Trial conditions differed with regard to both founder inset distance and offspring and pollen dispersal distances as indicated (see summaries for A through F to the right of the graph). Connected data points show alleles retained under one set of trial dispersal conditions when founders were placed at diagonally increasing distances from the preserve corner (x-axis). See text for more details.

Observed heterozygosity values for all trials after 100 mating bouts did not vary by a large amount (ranging from 0.967 to 0.980), showing only a 1.3% difference between the highest and lowest values. At generation 100 there was little population subdivision and inbreeding seen across trials with differing founder inset distances and various dispersal conditions (Figure 6). The highest F values were seen for populations under basic dispersal conditions (trial conditions A), followed by populations with offspring being dispersed less distantly (trial conditions B and C). Populations tended to have slightly greater F values at inset distances beyond 100 grid points, except for trial D, where offspring were more distantly dispersed (80% to the last frame) and trial F, where pollen was more distantly dispersed and offspring least distantly dispersed (80% from the last frame and 90% to the first frame, respectively).

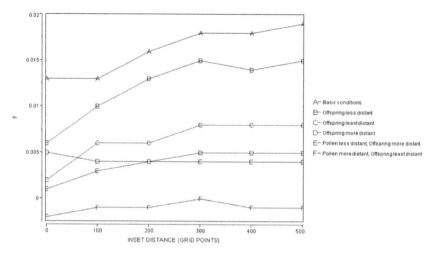

Fig. 6. Changes in F values in populations after 100 bouts of mating (the same populations shown in Figures 4 and 5). Trial conditions differed with regard to both founder inset distance and offspring and pollen dispersal distances as indicated (see summaries for A through F to the right of the graph). Connected data points show F values under one set of trial dispersal conditions when founders were placed at diagonally increasing distances from the preserve corner (x-axis). See text for more details.

4. Discussion

4.1 Practical applications

The null hypothesis that placing founders at differing distances from the edge of a restoration preserve will have no effect on subsequent population growth or genetic diversity was falsified according to results involving the first series trials with the basic dispersal conditions only (Figures 1 through 3). These results confirm that populations can manifest different degrees of population growth and genetic edge effects (Rogstad & Pelikan, 2011) depending on the distance of founders from preserve borders. In terms of population development, increases in growth rates were correlated with increasing distances of founder placement from a border up to a point: with the "basic" gene dispersal distances, planting founders 300 grid points (1500 m) into the preserve is as good as planting them 500 grid points (2500 m) into a 5 km x 5 km or larger preserve. Increasing founder placement up to 300 grid points increasingly allows founders a greater number of viable grid points to establish their offspring thereby reducing mortality levels. As for genetic diversity, the greatest loss of alleles occurred when founders were placed less than 100 grid units from the border (trial a, Figure 2). However, even at the corner, losses were not considerably greater than placing the founders 100 grid points into the preserve (corner trial a retained 90.4% while trial b retained 93.9% of the founding alleles). A preserve manager not interested in maximizing population growth but who only wants to maximize genetic diversity retention could thus plant the founders 100 grid points (500 m) into a 5 km x 5 km or larger preserve. Beyond that, planting further into the preserve does not significantly increase the number of founding alleles conserved. Differences in losses of

Founder Placement and Gene Dispersal Affect
Population Growth and Genetic Diversity in Restoration Plantings of American Chestnut

143

genetic diversity as indicated by changes in observed heterozygosity or F values were not pronounced at age 100 among these trials, and thus are not of major concern to a restoration manager through the initial establishment period modeled.

The second series of trials in which offspring and pollen movement distances were varied in populations with founders placed at increasing distances from the border (Figures 4 through 6) demonstrate that population and genetic edge effects are further affected by variations in gene dispersal distances. Compared to the basic conditions used in Figures 1 through 3, population growth was increasingly reduced (Figure 4) when pollen dispersal was more restricted and offspring dispersal was more distant (trial E), when offspring dispersal was less distant but pollen dispersal matched the basic conditions (trial B), when pollen dispersal was more distant and offspring dispersal less distant (trial F), and when offspring dispersal was most restricted but pollen dispersal matched the basic conditions (trial C). Only when offspring establishment was more distant but pollen dispersal matched the basic conditions (trial D) did population growth exceed that exhibited when the basic dispersal condition applied (trial A). These results suggest that in restoration projects with conditions similar to those simulated here, large gains in population growth might be promoted by ensuring that more offspring establish at greater distances than are occurring naturally.

Differences in founding allele retention among this second series of trials were also evident (Figure 5), with the greatest retention in alleles being under the basic and offspring more distant dispersal conditions (trials A and D respectively). For trial conditions A, B, C, and F, there was a sharp increase in allele retention by moving founders from the corner inwards 100 m, after which further gains were not as pronounced to various degrees for different trial conditions at increasing distances. Under trial conditions E and D, placing founders at 300 grid points would be preferable since significant gains in allele retention were not had beyond that distance. Across all of these trials, the greatest difference in allele retention was 9.6% (between trial E founders at the corner and trial D founders inset by 400 grid points). Inbreeding and subdivision appear to attain the highest values under the basic dispersal conditions with founders inset into the preserve (Figure 6), although none of the values are yet approaching pronounced levels. Causes driving the minor differences seen in F values among trials are not always readily interpretable (e.g., note trials E versus F versus C).

Overall, the relationships among trial conditions, inset distance, population growth, genetic diversity retention, and inbreeding/subdivision can be complex and are not necessarily intuitive. Given this complexity, NEWGARDEN modeling can be used to suggest restoration management strategies. As noted earlier, restoration programs and management need to be as cost-efficient as possible. By understanding interactions between life-history characteristics of C. dentata and inset distance for founder establishment, program managers can estimate the best methods to minimize such costs. For example, results of this study suggest that planting and stewarding a limited number of founders (thousands at one location are not needed) at least 1500 m into a preserve and promoting successful offspring establishment beyond that which is occurring naturally, rather than expending that effort promoting establishment within the developing stand, is less costly and potentially more successful in terms of ease of population establishment and growth, maintenance, as well as retention of genetic diversity and avoidance of inbreeding and subdivision compared to other options. Managers may have to make cost-benefit decisions such as which is more important: saving funds by planting near a border, maintaining genetic diversity at the sacrifice of maximizing population growth, or planting further into a preserve at greater expense. The results here stress the need for accurate knowledge of realized gene dispersal attributes (and of other life history attributes) in any modeling-derived restoration management planning. Lacking such information, restoration managers should preferably take the most conservative approach to restoration using as accurate estimations of life history characteristics as possible, in this case, planting the founders at least 1500 m from a border.

4.2 Evolutionary implications

Natural populations establish with varying numbers of founders (often low numbers), degrees of isolation, and geometry of founding. Interpopulation and interspecific life history variation compound the complexity of establishment events as indicated in trials here. The results above suggest that these factors can interact differentially among newly founded populations, and that even seemingly slight differences in initial conditions (e.g., just 500 m difference in founder placement) may have significant effects on the future trajectories of population growth and genetic diversity measures. NEWGARDEN can be used to explore this variation in theoretical and existing situations.

4.3 Further considerations

We emphasize that the results reported here were generated with a subset of the potential input conditions including constant input parameter values designed to simulate population development in the only known restoration population of chestnuts, the West Salem stand (Rogstad & Pelikan, 2011). However, conditions at other restoration sites will surely differ (e.g., density of competitors and rates-distances of establishment, age-specific offspring and pollen production, age-specific mortality rates, gene dispersal distances, etc.). Further, other species will often have drastically different conditions from those modeled here (e.g., short lived perennials or annuals, different densities, different gene dispersal patterns). Finally, the ecological niche will vary across species in ways that might affect geometric patterning (e.g., the section of a forest the species naturally inhabits: one species that would typically be found living on the outer regions of a forest would have different life history characteristics than a species that would inhabit the center part of that forest). NEWGARDEN can be used to explore such intra- and interspecific differences in the conditions of population establishment and development.

Previous studies (Rogstad & Pelikan, 2011) have indicated that differences in the geometric placement of founders (e.g., spacing between individuals; arranged in lines versus squares; founder subdivision into a series of smaller squares) can affect rates of population growth and genetic diversity maintenance. When planting a founding population, the simplest pattern is to plant trees in straight lines to make rectangular or square founder areas. In our experiments, the founders were established in a 13 x 13 square of individuals to be easily manipulated for each trial of varying inset distances. This square shape only allowed for one offspring to establish within the founding square between founding parents, which may not allow for higher rates of population growth in the first few years of reproduction if the offspring have limited dispersal. One question not addressed in our trials was whether or not varying geometry of founder establishment patterns (e.g., completely random founder locations, X-shaped lines, various straight lines, circles, or hollow squares) would affect

Founder Placement and Gene Dispersal Affect
Population Growth and Genetic Diversity in Restoration Plantings of American Chestnut

145

population growth and genetic diversity. Different establishment patterns may allow for a greater number of offspring to land in unoccupied grid points than in the square pattern used in the above trials, and possibly reduce the loss of genetic diversity. It is also possible that the geometric pattern of the founder establishment interacting with the inset distance would affect the overall population growth and maintenance of genetic diversity. Trial comparisons where geometry is varied in combination with distance from edges and dispersal distances are underway.

For the above trials, 169 founders were used based on previous studies (Lawrence *et al.*, 1995; Chakraborty, 1993) that suggest a minimum of approximately 172 founders is needed to capture the greatest majority of alleles from the source population while also minimizing reintroduction costs in restoration projects. NEWGARDEN could be used to further investigate what would be the minimal number of founders needed to minimize allele loss over generations under a range of conditions.

Since it is not possible to investigate differences in all input conditions *a priori*, NEWGARDEN modeling can also be used *a posteriori* to generate simulated populations reflecting conditions and population growth outcomes for stands that have already been established. Such simulations can then be used to evaluate the degree to which population growth and genetic diversity preservation could be improved by supplementations or manipulations of actual populations (e.g., planting more individuals, altering gene exchange distances via increased seed distribution, reducing mortality rates at certain life stages, etc.).

If the restoration program includes harvesting offspring to redistribute within the preserve or at new locations, where should the offspring be collected from? Samples could be taken in the immediate area surrounding the entire founding population. However, this might not be optimal due to travel distance into the preserve. It would be better to collect offspring closer to one edge of the preserve, minimizing travel distance. However, perhaps the farther in one direction the offspring are from the founders, the greater the risk of loss of genetic diversity and increased offspring subdivision. One area for further study would be to analyze cohort (not population) growth, heterozygosity and F values, and loss of genetic diversity in subregions that are at varying distances from the founders. This would allow restoration managers to determine how close to an edge they could collect offspring that would still retain the greatest genetic diversity for new plantings. The results of such experiments would also provide information on the erosion of genetic diversity over space and time in entire populations and the degree to which erosion might be localized.

5. Acknowledgements

We thank the Departments of Biological Sciences, Environmental Studies, and Mathematical Sciences at the University of Cincinnati. We are grateful for funding from the Women In Science and Engineering Program, also at the University of Cincinnai. Specifically, we thank C. Daley, U. Ghia, and anonymous reviewers for their contributions.

6. References

Barakat, A., DiLoreto, D., Zhang, Y., Smith, C., Baier, K., Powell, W., Wheeler, N., Sederoff, R., & Carlson, J. (2009). Comparison of the transcriptomes of American chestnut

(*Castanea dentata*) and Chinese chestnut (*Castanea mollisima*) in response to the chestnut blight infection. *BMC Plant Biology*, Vol. 2009, No. 9, (May 2009), 513-958, 1471-2229

Chakraborty, R. (1993). A class of population genetic questions formulated as the generalized occupancy problem. *Genetics*, Vol. 134, No. 3, (July 1993), 953-958,1061- 4036

Diskin, M., Steiner, C., & Hebard, F. (2006). Recovery of American chestnut characteristics following hybridization and backcross breeding to restore blight-ravaged *Castanea dentata*. *Forest Ecology and Management*, Vol. 223, No. 1-3, (March 2006), 439-447, 0378-1127

Elliot, K., & Swank, T. (2008). Long-term changes in forest composition and diversity following early logging (1919-1923) and the decline of the American chestnut (*Castanea dentata*). *Plant Ecology*, Vol. 197, No. 2, (September 2007), 155-172, 1573-5052

Huang, H., Dane, F., & Kubisiak, T. (1998). Allozyme and RAPD analysis of the genetic diversity and geographic variation in wild populations of the American chestnut (Fagaceae). *American Journal of Botany*, Vol. 85, No. 7, (July 1998), 1013-1021, 1537-2197

Jacobs, D. (2007). Towards development of silvical strategies for forest restoration of American chestnut (*Castanea dentata*) using blight resistant hybrids. *Biological Conservation*, Vol. 137, No. 4, (July 2007), 497-506, 0006-3207

Jansen, P., Bongers, F., & Van der Meer, P. (2008). Is farther seed dispersal better? Spatial patterns of offspring mortality in three rainforest tree species with different dispersal abilities. *Ecography*, Vol. 31, No. 1, (February 2008), 43-52, 09067590

Lawrence, M., Marshall, D., Davies, P. (1995). Genetics of genetic conservation. I. Sample size when collecting germplasm. *Euphytica*, Vol. 84, No. 2, (June 1995), 89-99, 00142336

Merritt, D., & Dixon, K. (2011). Restoration Seed Banks – A Matter of scale. *Science*, Vol. 332, No. 6028, (April 2011), 424-425, 1095-9203

Nathan, R., Schurr, F., Spiegel, O., Steinitz, O., Trakhtenbort, A., & Tsoar, A. (2008). Mechanisms of long-distance seed dispersal. *Trends in Ecology and Evolution*, Vol. 23, No. 11, (November 2008), 638-647, 0169 -5347.

Paillet, F. (2002). Chestnut: history and ecology of a transformed species. *Journal of Biogeography*, Vol. 29, No. 10-11, (2002), 1517-1530, 1365-2699

Pierson, S., Keiffer, C., McCarthy, B., & Rogstad, S. (2007). Limited reintroduction does not always lead to rapid loss of genetic diversity: An example from the American chestnut (*Castanea dentata*: Fagaceae). *Restoration Ecology*, Vol. 15, No. 3, (September 2007), 420-429, 1526-100X

Primack, R. (May 1, 2010). *Essentials of Conservation Biology* (5th edition), Sinauer Associates, 978-0-87893-640-3, Sunderland MA

Primack, R., & Miao, S. (1992). Dispersal can limit local plant distribution.*Conservation Biology*, Vol. 6, No. 4. (December 1992), 513–519, 0888-8892

Rogstad, S., & Pelikan, S., (September 2, 2011). *Genetic Diversity in Establishing Plant Populations: Founder Number and Geometry* (1st edition), Science Publishers, 157808721X, New Hampshire.

Russel, E. (1987). Pre-blight distribution of *Castanea dentata* (Marsh.) Borkh. Bulletin *of the Torrey Botanical Club*, Vol. 114, No. 2, (April 1987), 183-190, 0040-9618

Stilwell, K., Wilbur, H., Werth, C., & Taylor, D. (2003). Heterozygote advantage in the American Chestnut, *Castanea dentata* (Fagaceae). *American Journal of Botany*, Vol. 90, No. 2, (February 2003), 207-213, 1537-2197

8

Characterisation of the Amaranth Genetic Resources in the Czech Gene Bank

Dagmar Janovská[1], Petra Hlásná Čepková[2] and Mária Džunková[3]
[1]Department of Gene Bank, Crop Research Institute
[2]Department of Crop Sciences and Agroforestry in Tropics and Subtropics
Institute of Tropics and Subtropics, Czech University of Life Sciences Prague
[3]Department of Genomics and Health,
Centre for Public Health Research (CSISP), Valencia
[1,2]Czech Republic
[3]Spain

1. Introduction

The human species depends on plants. These constitute the basis for food, supply most of our needs (including clothes and shelter) and are used in industry for manufacturing fuels, medicines, fibres, rubber and other products. However, the number of plants that humans use for food is minimal, compared to the number of species existing in nature. Only 30 crops, the most outstanding of which are rice, wheat and maize, provide 95% of the calories needed in the human diet (Jaramillo & Baena, 2002). However, agricultural biodiversity is in sharp decline due to the effects of modernisation, such as concentration on a few competitive species and changes in diets. Since the beginning of agriculture, the world's farmers have developed roughly 10 000 plant species for use in food and fodder production. Today, only 150 crops feed most of the world's population, and just 12 crops provide 80% of dietary energy from plants, with rice, wheat, maize and potato providing 60%. It is estimated that about three quarters of the genetic diversity found in agricultural crops have been lost over the past century, and this genetic erosion continues (EC, 2007).

Humans need to add to their diet those crops of high yield and quality that can adapt to environmental conditions and resist pests and diseases. Advantage must be taken of native and exotic species, with nutritional or industrial potential, or new varieties must be developed. Improving crops, however, requires reserves of genetic materials whose conservation, management and use have barely begun to receive the attention that they deserve. Humans take advantage of plant genetic resources in as much as they are useful to us, which means that we must understand them, and know how to manage, maintain and use them rationally (Jaramillo & Baena, 2002). Information on genetic diversity and relationships within and among crop species and their wild relatives is essential for the efficient utilization of plant genetic resource collections for the efficient explanation of taxonomic relationships (Chan & Sun, 1997; Drzewiecki et al., 2003).

Amaranthus L. is a genus from *Amaranthaceae* family probably originated in America. This genus contains approximately 70 species of worldwide distribution including pigweeds, waterhemps, and grain amaranths (Sauer, 1967). The origin of various species of cultivated amaranths is not easy to trace because wild ancestors are pantropical cosmopolitan weeds (Espitia-Rangel, 1994). For human consumption there are cultivated grain amaranths – *A.caudatus*, *A. cruentus* and *A. hypochondriacus* and vegetable amaranths – mainly *A.dubius*, *A.tricolor* and *A.cruentus*. Grain amaranths are crop species of New World origin; *A. caudatus* from Andean Peru and Ecuador, *A. cruentus* and *A. hypochondriacus* from Mexico and Central America (Sauer, 1950; Drzewiecki, 2001). Nowadays, the grain amaranths are cultivated from the temperate to tropical zone and the vegetable amaranths mainly in the South Africa and South Asia (Jarošova et al., 1997).

Amaranths are very promising crops. The main reasons could be content of protein, fat and active substances. The content of seed protein is in the range 13 – 18% with very good balanced amino acids. The lysine content is relatively high in the comparison with common cereals. The content of crude proteins in leaves is from 27 to 49% in d.m. what is more than in the leaves in the spinach (Segura-Nieto, 1994). Amaranths have comparable or higher amounts of essential amino acids as whole egg protein (Drzewiecki et al., 2003). The fat content is in the range 0.8-8.0%. The linoleic acid is the predominant fatty acid, with lesser amount of oleic and palmitic acids. The oil also contains squalene, precursor of cholesterol, which is used in the cosmetics and as a penetrant and lubricant (Becker, 1994). Many compounds and extracts from amaranths possessed anti-diabetic, anti-hyperlipidemic, spermatogenic and anti-cholesterolemic effects (Sangameswaran & Jayakar, 2008; Girija et al., 2011), antioxidant and antimicrobial activity (Alvarez-Jubete et al., 2010; Tironi&Anon, 2010). Many consumers purchase amaranth because they want a wheat- and gluten-free product, like the nutritional profile of amaranth, or enjoy "exotic" foods in their diet (Brenner et al., 2000). Amaranth can be used also as a feed for pigs, hens, etc. (Pisarikova et al., 2005). From the cultivation point of view, amaranth is interesting for its heat and drought resistance and very low susceptibility to diseases and pests (Barba de la Rosa, 2009). Considering its agronomic importance, attention should be given to the cultivation, conservation, and sustainable utilization of this promising crop (Ray & Roy, 2009).

Unfortunately, amaranths are also very harmful weeds spread in all over the world. Weedy *Amaranthus* species (pigweeds) have been and continue to be a major problem in agronomic production. The weed amaranth *A. retroflexus* is considered one of the world's worst weeds. A major contributor to the noxious nature of these weedy species is their ability to efficiently adapt to the changes in agricultural management practices that are specifically designed to control and prevent colonization. For example, numerous populations of pigweeds have evolved herbicide resistance (Drzewiecki, 2001; Rayburn et al., 2005).

In the Czech Republic the cultivation of amaranth was introduced in the early 1990s (Michalova 1999; Moudry et al. 1999) and the collection of amaranth genetic resources was established in 1993 in the Czech Gene Bank. Due to the very positive effects on the human health, we try to find out genotypes suitable for the Czech conditions with utilization in the Czech cuisine. On the Czech market, there is very popular food made from amaranth flour such as chips, cookies, and breakfast cereals, etc. However, all amaranth seeds are imported into the Czech Republic from other countries. The demand for vegetable amaranth is also increasing. Presently, in the Gene Bank, there are stored 103 evaluated accessions. In the working collection (in the different stages of evaluations), there are more than 30 accessions. Seed samples of amaranth are obtained from other gene banks, universities, private subjects or from collecting missions from all over the world. It corresponds with international agreements and with The Czech National Programme on Conservation and Utilization of Plant Genetic Resources and Agro-biodiversity. For maintenance and utilization of plant genetic resources of amaranths, it is very important to know them from all sides. Genetic resources studies are oriented on evaluation of the most important biological characters, with respect to the effective utilization of genetic resources in breeding and agricultural practice. Good characterization and evaluation of genetic resources under conditions similar to those of their origin can provide breeders and users with valuable information on effective utilization of genetic resources for the breeding programmes and utilization. Characterization of genetic resources is focused mainly on morphological characters. The evaluation consists of data on plant growth and development, characteristics of plant stand, analysis of yield elements, etc. (Dotlačil et al., 2001). First steps of evaluations after seed samples receiving, are field evaluations. The phenological and morphological evaluation such as length of vegetation, plant height, length of inflorescence, colour of inflorescence, type of inflorescence, etc., is performed during vegetation. The length of vegetation is very important for amaranth cultivation in the Czech Republic, because many of the amaranths genotypes are sensitive to day-length. They remain in the vegetative period for a long time and create seeds after day-shortening (NRC, 1984). In the Czech Republic, they flower in the second half of September. Because the early frost, they cannot mature their seeds.

For genetic improvement of *Amaranthus*, germplasm collections will play a key role as well. However, only limited information is available on intra- and inter-specific genetic diversity and relationships within *Amaranthus* germplasm collections (Chan & Sun, 1997). In spite of the fact that it has been the object of many studies, the genus *Amaranthus* is still poorly understood, being widely considered as a "difficult" genus. Currently, the taxonomic problems are far from being clarified especially because of the widespread nomenclatural disorder caused chiefly by repeated misapplication of names (Costea et al., 2001) which is shown in Table 1. Due to variation of morphological characters, accurate classification of amaranth genetic resources is not always possible (Transue et al., 1994).

For preliminary identification of *Amaranthus* species, the useful tool can be the number, thickness, orientation and density of branches in inflorescences. The flowers are arranged in small and very contracted cymes, which are agglomerated, axillary and additionally arranged in racemose or spiciform terminal, large and complex synflorescences. Although extremely variable, there is usually a tendency towards a morphological "type" (Costea et al., 2001).

The colour of the seeds is commonly dark-brown to blackish, or whitish-yellowish, sometimes with reddish nuances at the species cultivated as cereals. Many cultivars of *A. caudatus* have pink cotyledons visible through the seed coat. The colour may be uniform or not, in the last case usually with the marginal zone paler. Weedy species and species used as a vegetable have mostly black or dark seeds (Costea et al., 2001; Jarošová et al., 1997; Das, 2011).

Many species of the genus are greatly affected by environmental factors (nutritional elements, water availability, light conditions, injurious factors, etc. exhibiting a great morphological variability with little taxonomic significance (Costea et al., 2001). All the above mentioned characteristics are useful for the taxonomy of the genus but difficult to use for the current identification of taxa (Costea et al., 2001). Also it is dependent on the cultivation in the field conditions. In the case of a gene bank, when seed samples are received, it is necessary to sow them in the field conditions for the morphological and phenological evaluations. But in the case of weedy species, it would be better to know, if the samples are not harmful weeds. We need to exclude weeds from our collection.

Many different methods of identification have been used for evaluation of amaranth diversity. RAPD analysis was successful in the investigation of the relationships of four *A. hypochondriacus* varieties (Barba de la Rosa et al., 2009). AFLP markers were successfully used to determine species what demonstrated taxonomic ambiguity at the basic morphologic level (Costea et al., 2006). Other methods such as ITS, ISSR and isozyme profile were used to get exhaustive view of interrelationship and relative closeness among amaranth species (Das, 2011; Xu & Sun, 2001). Also other methods such as electrophoresis profiles of proteins have been successfully used to clarify the taxonomy of many families. There was published, that electrophoresis can also be used to characterize the seed protein profiles of species and cultivars, compare cultivars of different geographical origin, and provide taxonomically useful descriptors that are substantially free from environmental influence. This method is rapid, relatively cheap, largely unaffected by the growth environment and eliminate to grow plant to maturity (Juan et al., 2007; Jugran et al., 2010). Drzewiecki (2001) used SDS PAGE of urea-soluble proteins of amaranth seeds for distinguishing both – species and their cultivars. Samples of seven species were divided into three groups by protein patterns according to similarity. According to solubility, Osborne (1907) divided proteins into four classes: albumins soluble in water, globulins soluble in high salt concentration, prolamins soluble in aqueous alcohol and glutelins soluble in acid or alkaline solutions (Segura-Nieto et al., 1994). The division into four protein fractions brings the possibility to see the differences among seed samples more clearly. The first general characterization of the protein fraction spectra of amaranth species was performed by Gorinstein et al. (1991) and Drzewiecki et al. (2003). Finally, Dzunkova et al. (2011) set up the methodology for clear identification of the amaranth species using glutelin protein fraction. The washing off water, salt- and alcohol- soluble proteins in protein fraction separation process makes polymorphic peaks of amaranth glutelins to be distinguished very easily.

SDS PAGE has been the traditional method for analysing glutenin subunit composition of wheat, but the procedure is slow, laborious and non-quantitative. The chip microfluidic technology, based on capillary electrophoresis, provides new opportunities in analysis of wheat HMW-GSs. This procedure is rapid, simple to operate, enabling automatic and immediate quantitative interpretation. Other advantages over traditional gel electrophoresis are lower sample and reagent volume requirements and a reduced exposure to hazardous chemicals (Bradova & Matejova, 2008).

In this work, we focused on evaluation for precise determination of amaranth genetic resources in the Czech Gene Bank. One of our aims was to separate amaranth species according to protein patterns and to verify our hypothesis of different protein fraction pattern based on species and variety. We compared spectra of storage proteins and their fractions of wild weedy and cultivated species of amaranths and verified the suitability of this method for species identification in our collection.

Latin name	Synonyms
Amaranthus caudatus L.[1]	*Amaranthus caudatus* subsp. *caudatus*
	Amaranthus caudatus subsp. *mantegazzianus*
Amaranthus caudatus subsp. *caudatus*[1]	= *Amaranthus alopecurus* Hochst. ex A. Br. & Bouche
	= *Amaranthus abyssinicus* hort. ex L.H. Bailey.
	= *Amaranthus caudatus* subsp. *saueri* Jehlik
	= *Amaranthus caudatus* L.
	= *Amaranthus maximus* Mill.
	= *Chenopodium millmi* J.T. del Granado
	= *Amaranthus caudatus* var. *alopecurus* Moq.
Amaranthus caudatus subsp. *mantegazzianus*[1]	= *Amaranthus edulis* Spegazz.
	= *Amaranthus mantegazzianus* Passer.
Amaranthus cruentus L. [1]	= *Amaranthus caudatus*auct.
	= *Amaranthus paniculatus* L.
	= *Amaranthus hybridus* var. *cruentus*
	= *Amaranthus sanguineus* L.
	= *Amaranthus hybridus* 'paniculatus'.
	= *Amaranthus speciosus*
Amaranthus retroflexus L. [1]	= *Amaranthus patulus*auct.
	= *Amaranthus delilei* Richter & Loret
Amaranthus hypochondriacus L. [1]	= *Amaranthus chlorostachys* var. *erythrostachys* (Moq.) Aell.
	= *Amaranthus leucospermus* S. Wats.
	= *Amaranthus leucocarpus* S. Wats.
	= *Amaranthus hybridus* convar. *erythrostachys* (Moq.) Thell. ex Asch. & Graebn.
	= *Amaranthus hybridus* subsp. *hypochondriacus* (L.) Thell.
	= *Amaranthus flavus* L.
	= *Amaranthus frumentacea* Buch.-Ham.
	= *Amaranthus chlorostachys* var. *leucocarpus* (S. Wats.) Aell.
	= *Amaranthus anardana* Buch.-Ham.
Amaranthus cannabinus (L.) J.D.Sauer[2]	*Acnida cannabina* L.
Amaranthus deflexus L. [2]	
Amaranthus tuberculatus (Moq.) J.D.Sauer[2]	*Acnida tuberculata* Moq.

[1]according to Mansfeld'sEncyclopedia of Agricultural and Horticultural Crops (Hanelt& IPGCPR, 2001)
[2]according to IPNI (2011)

Table 1. Synonyms of selected amaranth species

2. Materials and methods

2.1 Plant material

For the evaluation there were used 46 amaranth genotypes from Crop Research Institute in Prague, Czech Republic (CRI) and from USDA, ARS, NCRPIS Iowa State University. In these samples, there were 6 accessions of wild weed and 40 of the cultivated species. The acronyms used for the wild species were as follows: De - *A. deflexus*, Au - *A. australis*, Wr - *A. wrightii*, Tu - *A. tuberculatus*, Cn - *A. cannabinus*, Re - *A. retroflexus*. The cultivated samples were evaluated in the field conditions in 2008 and 2009 according to the list of descriptors for amaranths created for purposes of the Czech Gene Bank. The morphological and phenological characters are evaluated according to List of Descriptors for amaranth created in the Czech Gene Bank. Following traits were evaluated in the field conditions:

* number of days from emergence to inflorescence observation,
* number of days from emergence to flowering,
* number of days from emergence to maturity.

The first two traits were assessed when 50% of plants were in this stage. The numbers of day from emergence to inflorescence observation and the numbers of days from emergence to flowering are important characters due to fact, that certain amaranth genotypes are sensitive to day-length. Maturity was estimated when 75% of the grains were mature. Plant height was measured from the soil surface to the top of the main stem in cm. Length of inflorescence was measured from the downmost branch to the top of inflorescence of the main stem in cm. Weight of thousand seeds (WTS) was weight of thousand seeds in g.

2.2 Total seed protein content and protein fractions content determination

The measurements of total seed protein content and protein fraction content were performed by the Kjeldahl method (Czech state norm 56 0512-12) in Kjeltec automatic analyzer (Kjeltec 2300, Foss Tecator, Sweden) with the protein-nitrogen coefficient set to 6.025. Protein fractions (albumins, globulins, prolamins and total glutelins) were extracted according to the protocol developed for the wheat protein fraction separation by Dvoracek (2006) with some modifications. For the determination of protein fractions content was used 0.5 g of milled amaranth seeds. The protein fractions were extracted by adding 5 ml of solvent (distilled water for albumins, 0.5 M NaCl for globulins, cold 60% ethanol for prolamins), vortexing and centrifuging by $10\ 000 \times g$ for 15 minutes (Universal 32R HettichCentrifugen, Germany). This procedure was repeated twice and the supernatants from each extraction were saved and poured together. In the case of prolamins, after first addition of solvent, tubes were vortexed and chilled to 4°C for 4 hours; after that the procedure was performed exactly as for albumins and globulins. The protein content of whole seed was also measured by milling 1g of amaranth seeds. For the boiling in the automatic digestion system (2015lift, 2020 digestor, Foss Tecator, Denmark) were used 10 ml from the obtained 15 ml of each fraction extract. Into the each 250 ml tube one catalyser tablet, 3.5 g of K_2SO_4 and $CuSO_4$ mixture and 10 ml of H_2SO_4 were added. In one tube was a blank sample. Tubes were let to boil to the temperature of 420°C for about 1 h 40 min. After cooling for about 10 min, 75 ml of distilled water was added. The content of glutelin and the residual nitrogen fraction was calculated as the difference between the content of the total seed protein and three measured fractions.

2.3 Electrophoresis of the proteins

2.3.1 Extraction of the total seed storage proteins

Five different approaches to the extraction were tested for the development of the best extraction approach:

1. single seed extracted in 18 µl of the extraction solution,
2. bulk of 10 seeds extracted in 50 µl of the extraction solution,
3. bulk of 10 seeds extracted in 100 µl of the extraction solution,
4. bulk of 100 seeds extracted in 200 µl of the extraction solution,
5. bulk of 100 seeds extracted in 400 µl of the extraction solution.

Seed samples were crushed separately and mixed with extraction solution (consisted of 0.0625 M Tris-HCl pH 8.8, 5% (w/v) 2- mercaptoethanol, 2% (w/v) SDS, 10% (w/v) glycerol, 0.01% (w/v) bromphenol blue) by   vortexing (MS2 Minishaker, IKA, Germany) several times in 1.5 ml tubes. Tubes were allowed to stand at 4 °C for three hours. After this extraction time, the tubes were centrifuged at 12 000 x g for 15 min (Universal 32R HettichCentrifugen, Germany). After the replacement of the samples to the new tubes, the samples were heated in a boiling water bath for 2 min.

Ten seeds from each variety were selected randomly, crushed and put into 2 ml micro tube. The protein fractions were extracted by adding 100 µl of solvent (distilled water for albumins, 0.5 M NaCl for globulins, cold 60% ethanol for prolamins), vortexing and centrifuging by 10 000 × g for 15 minutes (Universal 32R HettichCentrifugen, Germany). This procedure was repeated twice but the supernatants of the second and third wash were always discarded. In the case of prolamins, after first addition of solvent, tubes were vortexed and chilled to 4°C for 4 hours; after that the procedure was performed as in the case of albumins and globulins. Tubes containing protein fractions extract and the seed pellets (glutelins) were frozen to -25°C. After the supernatant in the tubes became solid, the top of the tubes was perforated by a needle to form small holes what serve to prevent the loss of the sample by lyofilisation. The lyofilisation was performed by freeze dryer (Christ, Germany) during 24 h at -58°C and 0.018 mBar. The lyophilized solid samples were mixed with 100 µl extraction solution (consisted of 0.0625 M Tris-HCl pH 8.8, 5% (w/v) 2- mercaptoethanol, 2% (w/v) SDS, 10% (w/v) glycerol, 0.01% (w/v) bromphenol blue) by vortexing several times in 1.5 ml tubes. Tubes were allowed to stand at 4 °C for three hours. After this extraction time, the tubes were centrifuged at 12000 x g for 15 min. The supernatants were put into new tubes and heated in boiling water for 2 min.

2.3.2 Protein separation by SDS PAGE

The amaranth protein extracts were separated in conditions of discontinuous electrophoresis (SDS-PAGE) according to Laemmli (1970) 4% stacking gel of pH 6.8, 10% separation gel of pH 8.8 on the polyacrylamide gels of the size 180 x 160 x 0.75 mm.

On the gel was loaded:

-   15 µl of the single seed sample,
-   20 µl of the 10 seed bulk, 100 seed bulk and all the protein fraction samples,
-   7 µl of the protein marker: SigmaMarker Wide Range (MW 6,500-200,000).

The electrophoresis was performed on 90 mA (45 mA / gel) and let to run for about 4 hours. The gels were stained with a solution of 0.1% (w/v) Coomasie Brilliant Blue (CBB) R250, 50% (w/v) methanol, 10% acetic acid, 0.02% (w/v) bromphenol blue salt for 1 day and destained with a solution of 25% (w/v) denatured alcohol and 3.5% (w/v) acetic acid, what lasted also 1 day. Gels were preserved in solution: 45% (w/v) denatured alcohol, 3% (w/v) glycerol for 2 hours, then dried and stored into cellophane sheets. The whole procedure including the test of the different extraction concentrations, the protein fraction separation procedure and the electrophoresis was repeated for the control of the correct experiment performance.

2.3.3 Chip electrophoresis

All the extracted protein fraction samples were analyzed by chip capillary electrophoresis using commercial Experion Pro260 Analysis Kit for 10 Chips and the Experion automated electrophoresis system (Bio-Rad Laboratories, USA) for protein quantification according to the manufacturer's instructions. Experion automated electrophoresis station performs automatically all the steps of the gel-based electrophoresis (samples separation, staining, destaining, imaging, band detections, and data analysis).

2.4 Statistical analysis

For the statistical evaluation of morphological traits, analysis of variance (ANOVA) and the Tukey HSD test were used (software -Statistica 7.0 CZ). In the case of protein fraction proportion in accessions with different seed colour, the basic statistics of R statistics 2.10.0 software were used for calculation of mean x, standard deviation sx and p-values (adjusted by Holm correction, two sided Welch Two Sample t-test used).

The SDS-PAGE spectra of total seed storage proteins and protein fractions were compared and confronted with the spectra of the chip capillary electrophoresis. The bands in the spectra were analyzed regarding the positions of the bands and also the relative intensity of the bands. The intensity of the bands was analyzed individually for each sample considering the intensity of the internal markers of the chip electrophoresis and the general intensity of all the bands in the sample. The intensity of the bands was expressed as the relative protein concentration measured by chip capillary electrophoresis what was the multiplication of numbers 0, 1, 2, 3 used in our statistics (0- no band, 1- light band, 2 – medium intensity band, 3 – dark band). The spectra expressed as the numerical values were analyzed by R statistics 2.10.0 software. The relationships between accessions were expressed by Pearson correlation using single linkage. The hierarchical clustering dendrogram was cut at the level of correlation 0.99 to show the well defined clusters.

3. Results and Discussion

3.1 Morphological and phenological evaluation

Mean data of morphological and phenological evaluations of amaranth are shown in table 2 and 3. From our long-term observations, genotypes with number of days from emergence to flowering higher than 100 days likely does not mature before early frost in autumn. The vegetation period in evaluated collection ranged from 92±0.00 to 163.00±0.00 days. Also height of plants in maturity and length of inflorescence is a very useful character. Both are

Genotype	From emergence to inflorescence observation (days) Mean±SD	From emergence to flowering (days) Mean±SD	From emergence to maturity (days) Mean±SD
6	56.00±1.41[abc]	72.00±7.07[bcdef]	107.00±16.97[ab]
11	51.50±7.78[abc]	67.00±4.24[abcdef]	120.50±7.78[ab]
12	47.50±4.50[abc]	64.00±11.37[abcdef]	101.00±16.97[ab]
21	46.50±3.54[abc]	64.00±8.49[abcdef]	108.50±24.75[ab]
23	51.00±2.83[abc]	64.00±9.90[abcdef]	109.00±25.46[ab]
24	46.00±0.00[abc]	64.00±9.90[abcdef]	102.00±25.46[ab]
35	51.50±2.12[abc]	64.50±7.78[abcdef]	100.50±26.16[ab]
43	64.00±0.00[abc]	86.00±0.00[ef]	122.00±0.00[ab]
44	62.00±5.66[abc]	105.00±0.00[abcdef]	131.00±36.77[ab]
45	44.00±0.00[abc]	63.00±0.00[abcdef]	145.00±0.00[ab]
51	49.00±4.24[abc]	67.00±5.66[abcdef]	107.00±26.87[ab]
62	45.00±1.41[abc]	64.50±9.19[abcdef]	126.00±0.00[ab]
70	52.00±1.41[abc]	68.50±3.54[abcdef]	112.50±19.09[ab]
71	45.50±4.50[abc]	65.00±4.24[abcdef]	111.50±17.68[ab]
72	50.00±7.07[abc]	68.00±2.83[abcdef]	100.00±26.87[ab]
73	41.00±0.00[ab]	56.50±0.71[abcd]	113.50±0.71[ab]
75	52.00±2.83[abc]	75.00±0.00[cdef]	104.50±16.26[ab]
76	47.50±0.71[abc]	61.00±4.24[abcdef]	114.50±3.54[ab]
80	44.00±0.00[abc]	69.00±0.00[bcdef]	124.00±0.00[ab]
92	75.00±7.07[abc]	98.00±0.00[abcdef]	163.00±0.00[b]
95	34.50±19.09[ab]	51.00±7.07[abcd]	111.00±1.41[ab]
96	36.50±12.02[ab]	60.50±10.61[abcdef]	106.00±16.97[ab]
98	47.50±12.02[abc]	78.00±0.00[def]	121.00±0.00[ab]
99	45.00±0.00[abc]	63.00±0.00[abcdef]	111.00±0.00[ab]
101	42.50±0.71[abc]	55.50±3.54[abcd]	110.00±1.41[ab]
104	43.50±7.78[abc]	54.00±4.24[abcd]	114.50±20.51[ab]
107	45.00±0.00[abc]	65.00±0.00[abcdef]	111.00±0.00[ab]
109	48.00±0.00[abc]	88.00±0.00[f]	101.50±14.85[ab]
110	49.50±2.12[abc]	64.50±4.50[abcdef]	118.00±11.31[ab]
111	52.50±2.12[abc]	63.00±1.41[abcdef]	116.50±9.19[ab]
112	55.50±7.78[abc]	70.50±10.61[bcdef]	120.00±14.14[ab]
120	51.50±0.71[abc]	70.00±0.00[bcdef]	116.00±22.63[ab]
121	31.50±23.33[ab]	57.00±5.66[abcd]	92.00±0.00[a]
123	41.50±14.85[ab]	53.00±15.56[abcd]	114.00±25.46[ab]
124	47.50±3.54[abc]	60.00±2.83[abcde]	115.00±11.31[ab]
125	49.5±0.71[abc]	58.00±0.00[abcd]	105.50±7.78[ab]
132	35.00±0.00[ab]	47.00±1.41[ab]	97.00±15.56[ab]
134	35.00±2.83[ab]	48.50±0.71[abc]	97.50±16.26[ab]
136	27.50±19.09[a]	41.00±16.97[a]	103.00±22.62[ab]
143	41.50±3.54[ab]	60.00±4.24[abcde]	98.50±17.68[ab]
Year			
2008	45.44±11.67[a]	66.21±13.60[a]	115.68±15.42[a]
2009	48.17±8.78[a]	63.83±13.06[a]	108.27±19.29[b]

SD-standard deviation
Analysis of variance (ANOVA) and the Tukey HSD test were used for statistical evaluation (software - Statistica 7.0 CZ).
Different letters in the same row are statistically significant at p > 0.05.

Table 2. Phenological evaluation of amaranths

Genotype	Inflorescence length (cm) Mean±SD	Plant height (cm) Mean±SD	WTS (g) Mean±SD	Colour of seed
6	54.00±1.41[de]	137.50±3.54[bcd]	0.75±0.01[cdefghijk]	pale
11	56.50±0.71[de]	137.50±45.96[bcd]	0.69±0.08[bcdefghij]	pale
12	29.50±0.71[bc]	117.50±3.54[abcd]	0.68±0.04[bcdefghij]	pale
21	34.00±1.41[bcd]	90.00±7.07[abc]	0.58±0.03[abcdefg]	pink
23	44.50±0.71[cde]	127.50±3.54[bcd]	0.76±0.08[cdefghijk]	black
24	66.00±1.41[e]	152.50±3.54[cd]	0.75±0.07[cdefghijk]	black
35	30.50±0.71[bc]	167.50±3.54[d]	0.90±0.00[ijk]	pale
43	36.00±0.00[bcd]	150.00±0.00[cd]	0.74±0.04[cdefghijk]	pale
44	24.50±0.71[a]	147.50±3.54[cd]	0.74±0.00[cdefghijk]	pale
45	29.00±0.00[bc]	100.00±0.00[abc]	0.88±0.00[hijk]	pale
51	29.00±0.00[bc]	142.50±3.54[bcd]	0.85±0.00[hijk]	pale
62	37.50±0.71[bcd]	137.50±3.54[bcd]	0.66±0.06[abcdefghi]	black
70	45.50±0.71[cde]	102.50±3.54[abcd]	0.78±0.05[efghijk]	pale
71	52.50±0.71[cde]	132.50±3.54[bcd]	0.86±0.15[hijk]	pale
72	46.50±0.71[cde]	122.50±3.54[bcd]	0.93±0.10[jk]	pale
73	41.50±0.71[cde]	130.00±8.49[bcd]	0.71±0.01[cdefghijk]	pale
75	60.00±0.00[e]	132.50±3.54[bcd]	0.91±0.10[ijk]	pale
76	36.50±3.54[bcd]	109.00±4.24[abcd]	0.84±0.08[ghijk]	pale
80	47.00±0.00[cde]	125.00±0.00[bcd]	0.50±0.00[abcd]	black
92	51.00±1.41[de]	142.50±3.54[bcd]	0.84±0.00[ghijk]	pale
95	35.50±0.71[bcd]	92.50±3.54[abc]	0.63±0.11[abcdefgh]	black
96	34.00±1.41[bcd]	92.60±3.54[abc]	0.40±0.04[a]	black
98	43.50±0.71[cde]	152.50±3.54[cd]	0.96±0.00[k]	pale
99	44.00±0.00[cde]	110.00±0.00[abcd]	0.70±0.00[bcdefghijk]	black
101	38.00±0.00[bcd]	127.50±36.06[bcd]	0.5±0.02[abc]	pink
104	42.50±0.71[cde]	92.50±3.54[abc]	0.63±0.04[abcdefgh]	black
107	22.00±0.00[a]	53.00±0.00[a]	0.52±0.00[abcde]	black
109	34.50±0.71[bcd]	77.50±3.54[ab]	0.44±0.04[ab]	black
110	49.50±0.71[cde]	155.00±7.07[cd]	0.70±0.00[bcdefghijk]	black
111	52.00±1.41[de]	137.50±3.54[bcd]	0.52±0.11[abcde]	black
112	53.00±2.83[de]	127.50±3.54[bcd]	0.84±0.01[ghijk]	pale
120	51.50±2.12[de]	95.00±42.43[abc]	0.73±0.03[cdefghijk]	pale
121	36.50±2.12[bcd]	115.00±21.21[abcd]	0.54±0.17[abcdef]	pale
123	47.00±0.00[cde]	102.50±3.54[abcd]	0.80±0.00[fghijk]	pale
124	54.50±0.71[de]	137.50±3.54[bcd]	0.82±0.06[ghijk]	pale
125	51.50±0.71[de]	127.50±3.54[bcd]	0.75±0.14[cdefghijk]	pale
132	46.50±4.50[cde]	114.50±23.33[abcd]	0.77±0.01[efghijk]	pale
134	44.50±3.54[cde]	119.50±17.68[bcd]	0.81±0.08[fghijk]	pale
136	45.50±3.54[cde]	92.50±38.89[abc]	0.77±0.01[defghijk]	pale
143	39.50±2.12[bcd]	107.50±38.89[abcd]	0.79±0.02[efghijk]	pale
Year				
2008	42.71±9.90[a]	122.85±26.21a	0.72±0.15[a]	
2009	43.22±9.87[a]	119.02±26.36[a]	0.73±0.14[a]	

Different letters in the same row are statistically significant at p > 0.05.

SD-standard deviation

Analysis of variance (ANOVA) and the Tukey HSD test were used for statistical evaluation (software - Statistica 7.0 CZ).

Table 3. Morphological evaluation of amaranth

important for mechanized harvest by combine harvester. Lower plants with mean inflorescence are better for grain production and mechanized harvest. From our collection it is for example accession '120' with 95.00±42.43 cm height and 51.50±2.12 cm length of inflorescence. Taller genotypes are useful to develop varieties for feed utilization (Wu eta l., 2000). On the other hand, plant height could be influenced by increasing of number of plant per m^2 (Jarošová et al., 1997). The value of weight of thousand seeds (WTS) is shown in table 3. In the relation with seed colour is clear, that the biggest WTS was observed in pale seeded samples. The seed size of the genera ranges from 0.37 to 1.21 g per 1000 seed weight according to Espitia-Rangel (1994). He noted that the low value corresponding to wild and weedy species and the high values to cultivated grain species. In our experiments the WTS ranged from 0.39 to 0.96 g.

3.2 Protein content and content of protein fractions

The results of the protein content analysis showed that the highest protein content (17.32 ± 0.82%) had *A. cruentus*accessions followed by *A. caudatus* (17.24±0.65%) and *A. hypochondriacus* (16.89±0.80%). It corresponds with other published data. Segura-Nieto et al. (1994) published, that the range of protein content is following: *A. cruentus* 13.2 – 18.2%, *A. hypochondriacus* 17.9% and *A. caudatus* 17.6 – 18.4%. The range of the total protein content into our collection (12.43 – 17.33%) was similar to the results of other authors investigating various amaranth genotypes (Barba de la Rosa et al., 2009). The amaranth albumins, globulins and prolamins formed 9.2 – 14.65%, 9.78 – 13.81% and 1.76 – 3.3% of total seed protein, respectively (Table 4). The glutelins with the residual nitrogen were the most abundant. It was in accordance with the results of Bressani & Garcia-Vela (1990) and Bejosano & Corke (1999a). The very low content of prolamins (1.76 – 3.3%) confirmed the results of several authors (Gorinstein et al., 1991a; Bejosano & Corke, 1999a; Petr et al., 2003). However, another group of authors reported several times more prolamins (Correa et al., 1986; Zheleznov et al., 1997; Vasco-Mendez & Paredes-Lopez, 1995). The differences between the results of these two groups of authors might be due to the different extraction methods (Fidantsi & Doxastakis, 2001). Significant differences between black, pale and pink coloured seeds in the content of albumins were detected. Content of albumins of the black seeded group (9.64 ± 0.40%) was significantly lower (p-value 4.10^{-3}) than of the pale seeded group (13.21 ± 1.45%) and also lower than of the pink seeded group (11.39 ± 0.00; p-value 2.10^{-2}). Bresani & Garcia-Vela (1990) did not observed any differences in the protein fractions distribution among species or cultivars of the same species, independent of the fractionation sequence used. However, our results showed that the black seeded varieties had the lowest albumin content. No significant differences in other protein fractions were detected.

	Seed colour			
	black	pale	pink	range
WTS (g)	0.60 ± 0.12	0.79 ± 0.09	0.54 ± 0.04	0.39 - 0.96
Protein content in %	15.69 ± 0.60	16.69 ± 0.78	16.04 ± 0.00	12.43 – 17.33
Albumins	9.64 ± 0.40	13.21± 1.45	11.39 ± 0.00	9.2 – 14.65
Globulins	10.92 ± 0.78 1	11.76 ± 1.72	10.75 ± 0.00	9.78 – 13.81
Prolamins	2.37 ± 0.82	2.68 ± 0.44	2.00 ± 0.00	1.76 – 3.3
Glutelins + residual nitrogen	77.07 ± 0.33	72.35 ± 2.67	75.86 ± 0.00	69.13 – 77.44

Table 4. Total seed protein content and protein fraction content (in % of DW) of investigated accessions with respect the seed colour.

3.3 Methodical approach to protein extraction

According to our results, the chip capillary electrophoresis could replace the standard SDS-PAGE procedure, because it produced comparable results and what is more it could be performed routinely also in small laboratories thanks to its rapid performance. On the other hand, the chip capillary electrophoresis showed wider range of proteins spectra (up to 260 kDa).

The test of different concentrations was used for selection of the best extraction approach for chip and SDS-PAGE electrophoresis. By the chip capillary electrophoresis, the bulked samples of 100 seeds in 400 µl of extraction buffer were also tested. The chip capillary electrophoresis showed the high sensitivity and therefore the high concentration of the protein in the main bands resulted in their illegility. The protocol of chip electrophoresis does not provide many possibilities to chase the loaded amount of the sample. The satisfactory results of the chip electrophoresis brought the use of the single seeds.

For the SDS PAGE there were used single seed samples, bulked samples of 10 seeds extracted in 50 and 100 µl and bulked samples of 100 seeds extracted in 200 and 400 µl of extraction solution were used. The protein patterns of the samples extracted from the single seeds did not show the intensity required for the analysis of all the bands in the spectra (Figure 1). On the other hand, samples obtained by extraction of 10 seeds in 50 µl and 100 seeds extracted in 200 µl of extraction solution did not show clearly separated bands, what resulted in their illegility. In comparison with the spectra of the less concentrated samples (single seeds, 10 seeds in 100 µl of extraction solution), the main bands of the more concentrated samples were thick and joined together. The bands, which were in the less concentrated samples less intensive, were expressed so intensively that formed dark background what resulted in the impossibility of identification of the individual bands in the protein spectra. The protein spectra of the samples obtained by the extraction of 100 seeds in 400 µl were also over expressed, but the less intensive bands did not form the background, so the mayor bands were more easily identified, but several mayor bands joined together.

As the best approach for the total seed storage protein extraction for classical SDS-PAGE we selected bulked samples of 10 seed extracted in 100 µl of extraction solution. The bulked samples of 10 seeds extracted in 100 µl to be the most suitable tools, because of their clear expression of protein patterns and moreover they can be used when samples with higher number of seeds are not available. This selected approach differed from methodology selected by Drzewiecki (2001) who used 50 µl or by Gorinstein et al. (2005) who used 62.5 µl of extraction solution for 10 seeds bulked samples. The need for using more extraction solution in our study might be to consequence of higher protein extraction as a result of the proper seed crushing performed in our study which was not mentioned in the methodology description of other authors (Drzewiecki, 2001; Gorinstein et al., 2005).

When using total seed storage protein spectra for accessions identification by chip electrophoresis the single seed samples with several repetitions showed up as the best approach. These results were with accordance with Bradova & Matejova (2008) that compared whole seed storage proteins of wheat.

3.4 Polymorphism of the glutelins

The electrophoresis of the glutelin fraction is widely used for crop varieties identification. There were published several articles about wheat (Matejova&Bradova, 2008; Dutta et al.,

2011), rice (Gorinstein et al., 2003), barley (Smith & Simpson, 1983), lupine (Vaz et al., 2004) etc. varieties identification based on glutelin patterns. Similarly amaranth glutelins showed polymorphism not only in position of bands but also in their intensity.

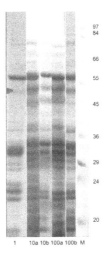

1 - a single seed extracted in 18 μl,
10a - bulk of 10 seeds extracted in 50 μl,
10b - bulk of 10 seeds extracted in 100 μl,
100b - bulk of 100 seeds extracted in 400 μl of the extraction solution.
M - wide range protein marker (bands in kDa).

Fig. 1. SDS – PAGE spectra of total seed storage proteins of sample obtained by different extraction approaches.

In the cluster dendrogram (Figure2), there were clearly separated the grain and the wild monoecious and the wild dioecious accessions. All investigated amaranth species had in common three major bands of the MW 21 – 23 kDa, but remarkable differences in the rest of the spectra were the reason for the segregation into three main clusters. The glutelin spectra of the grain amaranth varieties were very similar to the total seed storage protein patterns, but the main polymorphic bands were better distinguished because of the washing off the first three fractions during fraction separation procedure which probably formed the "background" of the spectra. The principal polymorphism was detected in following band positions 38, 39, 54, 58, 60, 64 and 65 kDa with three intensity levels (1-3). The amaranth glutelins showed up as the most abundant protein fraction by SDS-PAGE analysis also in the study of Bejosano&Corke (1999). The division of the grain amaranth glutelins into three major groups reported also Gorinstein et al. (2004) and Barba de la Rosa et al. (2009).

Figure 2 indicated three well defined clusters: grain species, monoecious wild species and dioecious wild species. The grain species *A. cruentus*, *A. hypochondriacus*, *A. caudatus* closely matched together with one sample *A. mantegazzianus*. There were clearly segregated clusters with the wild monoecious species (*A.wrightii*, *A. delfexus* and *A. retroflexus*) and the wild dioecous species (*A. australis*, *A. cannabinus* and *A. tuberculatus*).

A. caudatus group presentedwas two accessions '21' and '101' characterized by the dark band 60 kDa and the light band 39 kDa in their glutelin spectra. The *A. cruentus* cluster was clearly separated in the dendrogram of hierarchical distancing by the presence of the dark band of 58 kDa and of the light band in the position of 39 kDa. *A. hypochondriacus* accessions were characterized by the lack of any band in the position 58 kDa and by the presence of the dark band 54 kDa and the light band 38 kDa. The typical band (in the position 54 kDa) used for *A. hypochondriacus* recognition was qualified as characteristic for *A. hypochondriacus* by several authors (Drzewiecki, 2001; Marcone, 2002; Gorinstein et al., 2005), but its position was determined differently: as 55 kDa (Marcone, 2002) or 52 kDa (Drzewiecki, 2001) or in the case of protein fractions as 55 kDa, too (Thanapornpoonpong et al., 2008). The characteristic presence of the band 58 kDa in *A. cruentus* spectra and of the band 54 kDa in *A. hypochondriacus* spectra was confirmed by the results of Thanapornpoonpong et al. (2008).

Some of the accessions possessed extra light band of 65 kDa and were aggregated close to the *A. hypochondriacus* cluster. Their similarity to the other *A. hypochondriacus* varieties was expressed by very high correlation 0.987.

The dark band of 54 kDa, the dark band of 64 kDa and the light band in the position 65 kDa showed up in the glutelin spectra of the accession '134'. The accession '80' had the same glutelin spectra, but its band of 54 kDa was of medium intensity. These two varieties might be the hybrids of *A. hypochondriacus* and other unknown species which could have dark band of 64 kDa and light band of 39 kDa or they might be *A. hypochondriacus* varieties with some special properties that were not considered in our study. The accessions '132' with the dark band of 60 kDa typical for *A. caudatus* accessions was also present in the spectra and therefore the correlation between these accessions and the *A. caudatus* accessions was as high as 0.911. These accessions also showed the light band of 38 kDa and the medium intensity band of 54 kDa (typical marker for *A. hypochondriacus* spectra).

The dioecious wild species *A. australis*, *A. cannabinus* and *A. tuberculatus* formed a totally distinct cluster. They possessed several major dark bands of lower molecular weight 32 - 50 kDa. From this group, *A. cannabinus* and *A. australis* were the most similar, their correlation was 0.675. The monoecious wild species (*A. wrightii*, *A. deflexus* and *A. retroflexus*) and the dioecious wild species had in common one light band in the position of 65 kDa. The major dark bands of the monoecious wild species were of MW 29 - 66 kDa. The spectra of the monoecious wild species had some similarities with the spectra of the grain species. The grain species spectra were characterized by the two bands of MW 31 and 33 kDa while in the spectra of *A. retroflexus* these bands were just "shifted up" to MW 32 and 34 kDa. Protein fractions spectra of the wild species had not been published yet by other researchers. The results indicated the high correlation of the spectra of *A. retroflexus* and *A. wrightii* what confirmed the similarity observed by the first morphological descriptions made by Watson (1877).

Accessions possessing several bands of different intensities in the polymorphic area were qualified as the hybrid accessions. The accession '99' had in its spectra several bands in the polymorphic area: the dark band of 54 kDa, light band of 58 kDa, medium intensity band in the position of 60 kDa and the light band of 65 kDa. Its similarity with *A. hypochondriacus* was expressed as correlation 0.901 and to the accession '95'. The accession '95' differed from the accession '99' just in the intensity of the bands of 58 kDa and 60 kDa (correlation 0.971). Varieties '62' and '110' were designated as hybrid varieties. They had the both bands of 54 kDa (marker for *A. cruentus*) and 58 kDa (marker for *A. hypochondriacus*) of medium intensity. Moreover, they possessed the light band of 39 kDa. The presence of the light band 39 kDa (typical marker for *A. cruentus*) was the reason for their higher correlation with *A. cruentus* group (0.920) than with *A. hypochondriacus* group (0.892). The variety '111' was exceptional. Moreover, it had higher correlation with *A. hypochondriacus* varieties (0.960).

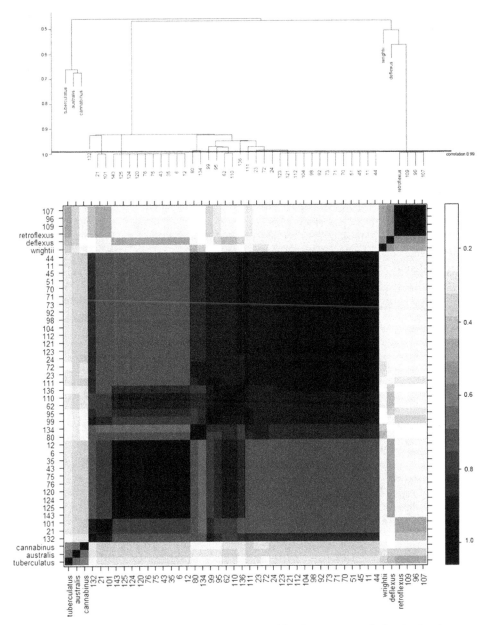

Fig. 2. Relations among amaranth samples expressed by Pearson correlation in dendrogram

4. Conclusion

Amaranth is mostly named as a crop of the future. Due to very good contents of protein, oil and many components with positive effects to humans, it is one of the promising crops. In the Czech Republic, there was interest of amaranth growing in the fields and the consumption of amaranth products is increasing as well. Most of grain raw material is imported to the Czech Republic from other countries, but there is increasing demand of Czech amaranth production. For amaranth cultivation it is necessary to know, what species could be grown. Because amaranth is not native in Europe, we have to receive seeds from other sides. In Czech legislation act about invasive weeds exists. Several amaranth species are included in this Act. In order to avoid cultivation of weedy amaranths, it is necessary to know the characteristics of the cultivated species and do not confuse them. Due to vegetable and weedy amaranth have black seed colour, it is impossible to use this trait as a marker. Amaranth glutelins were the best tool for the amaranth species identification, because they showed high polymorphism not only in position of bands but also in their intensity. The method used here was based on the data concerning the relative intensity and the position of the bands in the glutelin spectra obtained by the chip capillary electrophoresis what resulted in the exact similarity calculation of the protein fraction spectra and thus in the segregation of the cultivated grain species, the monoecious wild species and the dioecious wild species into three separate clusters. Each of the grain amaranth species was characterized by one dark band in the polymorphic region (54 – 65 kDa), while the hybrids possessed more bands of different relative intensity. The study brought several new contributions to the amaranth genetic research and is a very useful tool for species identification before cultivation in the field conditions. Unfortunately, this method is not so sensitive for individual amaranth genotype identification. We work on it in our current tasks.

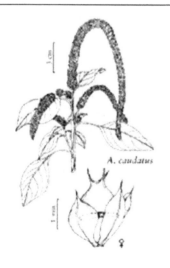

Fig. 3. *A. caudatus* (Standley, 1949)

Fig. 4. *A. hypochondriacus* (NRC, 1984)

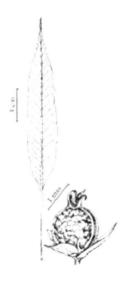

Fig. 5. *A. cruentus* (Bojian et al., 2003)

Fig. 6. *A. cannabinus* (Standley, 1949)

Fig. 7. *A. australis* (Standley, 1949)

Fig. 8. *A. retroflexus* (Standley, 1949)

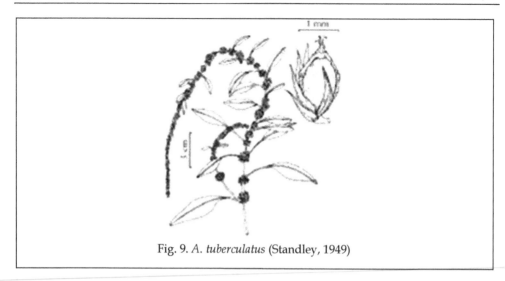

Fig. 9. *A. tuberculatus* (Standley, 1949)

5. Acknowledgement

This work was supported by the Czech Ministry of Agriculture QH92111.

6. References

Alvarez-Jubete, L.; Wijngaard, H.; Arendt, E.K. & Gallagher, E. (2010). Polyphenol composition and in vitro antioxidant activity of amaranth, quinoa, buckwheat and wheat as affected by sprouting and Baking. *Food Chemistry* 119, pp. 770–778, ISSN 0308-8146

Barba de la Rosa, A.P.; Fomsgaard, I. S.; Larsen, B.; Mortensen, A. G.; Olvera-Martınez, L.; Silva-Sanchez, C.; Mendoza-Herrera, A.; Gonzalez-Castaneda J. & De Leon-Rodrıguez A. (2009). Amaranth (*Amaranthus hypochondriacus*) as an alternative crop for sustainable food production: Phenolic acids and flavonoids with potentialimpact on its nutraceutical quality. *Journal of Cereal Science* 49, pp. 117–121, ISSN 0733-5210

Becker, R. (1994). Amaranth Oil: Composition, Processing, and Nutritional Qualities. In: *Amaranth: Biology, Chemistry and Technology*. Paredes-Lopez O. (Ed.) 133-141. CRC, ISBN 0-8493-5374-2, Boca Raton, USA

Bejosano, F. P. & Corke, H. (1999). Properties of protein concentrates and hydrolysates from *Amaranthus* and buckwheat. *Industrial Crops and Products*, 10 (3), pp. 175-183, ISSN 09266690

Bojian, B.; Clemants, S. E. & Borsch T. (2003). *Amaranthaceae*. Flora of China 9: 415-429.Available on internet http://hua.huh.harvard.edu/china/mss/welcome.htm [accessed 15. August 2011]

Brenner, D. M.; Baltensperger, D. D.; Kulakow, P. A.; Lehmann, J. W.; Myers, R. L.; Slabbert, M. M. & Sleugh, B. B. (2000). Genetic resources and breeding of Amaranthus. *Plant Breeding Reviews* 19, pp. 227-285. ISBN 0-471-38787-8, John Wiley & sons Inc.

Bressani, R. & Garcia-Vela, L. A. (1990). Protein Fractions in Amaranth Grain and Their Chemical Characterization. *Journal of Agricultural and Food Chemistry*, 38 (5), pp. 1205-1209, ISSN 0021-8561

Bradova, J. & Matejova, E. (2008).Comparison of the Results of SDS PAGE and Chip Electrophoresis of Wheat Storage Proteins. *Chromatographia* 67, pp. S83–S88, ISSN 0009-5893

Chan, K. F. & Sun, M. (1997). Genetic diversity and relationships detected by isozyme and RAPD analysis of crop and wild species of Amaranthus. *Theoretical and applied genetics.* 95: 865-873. ISSN: 1432-2242

Correa, A. D.; Jokl, L. & Carlsson, R. (1986). Amino-acid composition of some *Amaranthus* sp. grain proteins and of its fractions. *Archivos Latinoamericanos de Nutricion* 36 (3), pp. 466-476,ISSN 0004-0622

Costea, M.; Sanders, A. & Waines, G.J. (2001). Preliminary results towards revisions of the *Amaranthus hybridus* species complex (Amaranthaceae). SIDA 19 (4), pp. 931 – 974, ISSN 0833-1475

Das, S. (2011). Systematics and taxonomic delimitation of vegetable, grain and weed amaranths: a morphological and biochemical approach. *Genetic Resources and Crop Evolution*, 58, pp. 1-15, ISSN 1573-5109

Dotlačil, L.; Stehno, Z.; Michalová, A. & Faberová, I. (2001). Plant genetic resources and agri-biodiversity in the Czech Republic. In: *Agriculture and Biodiversity: Developing Indicators for Policy Analysis Proceedings from an OECD Expert Meeting Zurich*, November 2001. 66-79, Switzerland

Drzewiecki J. (2001). Similarities and Differences between *Amaranthus* Species and Cultivars and Estimation of Outcrossing Rate on the Basis of Electrophoretic Separations of Urea-Soluble Seed Proteins. *Euphytica*, 119 (3), pp. 279-287, ISSN 1573-5060

Drzewiecki, J.; Delgado-Licon, E.; Haruenkit, R.; Pawelzik, E.; Martin-Belloso, O.; Park, Y. S.; Jung, S. T.; Trakhtenberg, S. & Gorinstein, S. (2003). Identification and Differences of Total Proteins and Their Soluble Fractions in Some Pseudocereals Based on Electrophoretic Patterns. *Journal of Agricultural and Food Chemistry* 51 (26), pp. 7798-7804, ISSN 0021-8561

Dvoracek V. (2006). Optimalizace Osbornovy metody kvantifikace bílkovinných frakcí zrna pšenice ozimé (*Triticum aestivum* L.). 10p. ISBN 80-86555-81-X – Available on: http://www.vurv.cz/files/Publications/OptimalizaceOsborn.pdf [cited: 2.9.2011]

Dutta, T.; Kaur, H.; Singh, S.; Mishra, A.; Tripathi, J. K.; Singh, N.; Pareek, A. & Singh, P. (2011). Developmental changes in storage proteins and peptidylprolyl*cis*-*trans*isomerase activity in grains of different wheat cultivars. *Food Chemistry.* 12, pp. 450-457, ISSN 0308-8146

Dzunkova, M.; Janovska, D.; Hlasna-Cepkova, P.; Prohaskova, A. & Kolar, M. (2011). Glutelin protein fraction as a tool for clear identification of Amaranth accessions. *Journal of Cereal Science*, 53, 2, pp. 198 – 205, ISSN 0733-5210

EC (2007). *Genetic resources in agriculture. A summary of the projects co-financed under Council Regulation (EC) No 1467/94 Community programme 1994–99*, ISBN: 92-79-03599-1, Rome, Italy

Espitia-Rangel, E. (1994). Breeding of Grain Amaranth. In: *Amaranth: Biology, Chemistry and Technology.* Paredes-Lopez O. (Ed.).23-38. CRC, ISBN 0-8493-5374-2, Boca Raton, USA

Fidantsi, A. & Doxastakis, G. (2001). Emulsifying and foaming properties of amaranth seed protein isolates. *Colloids and surfaces B: Biointerfaces*, 21, 1-3, pp. 119-124, ISSN 1873-4367

Girija, K.; Lakshman, K.; Chandrika, U.; Ghosh, S. S. & Divya, T. (2011). Anti-diabetic and anti-cholesterolemic activity of methanol extracts of three species of Amaranthus. *Asian Pacific Journal of Tropical Biomedicine*, pp. 133-138, ISSN 2221-1691

Gorinstein, S.; Moshe R. & Greene, L. J. (1991). Evaluation of four *Amaranthus* species through protein electrophoretic patterns and their amino acid composition.*Journal of Agricultural and Food Chemistry*, 39 (5), pp. 851-854, ISSN 0021-8561

Gorinstein, S.; Pawelzik, E.; Delgado-Licon, E.; Haruenkit, R.; Weisz, M. & Traktenberg, S. (2002). Characterization of Pseudocereals and Cereals Proteins by Protein and Amino Acid Analyses. *Journal of the Science of the Food and Agriculture* 82 (8), pp. 886-891, ISSN 1097-0010

Gorinstein, S.; Yamamoto, K.; Kobayashi, S.; Taniguchi, H.; Pawelzik, E.; Delgado-Licon, E.; Shaoxian, Y.; Hongliang, S.; Martinez-Ayala, A. L. & Trakhtenberg, S. (2003). Inter-Relationship between Electrophoretic Characteristics of Pseudocereal and Cereal Proteins and Their Microscopic Structure for Possible Substitution Based on Nutritional Evaluation. *International Journal of Food Sciences and Nutrition* 54 (6), pp. 427- 435, ISSN 1465-3478

Gorinstein, S.; Pawelzik, E.; Delgado-Licon, E.; Yamamoto, K.; Kobayashi, S.; Taniguchi, H.; Haruenkit, R.; Park, Y.; Jung, S.; Drzewiecki, J. & Trakhtenberg, S. (2004). Use of scanning electron microscopy to indicate the similarities and differences in pseudocereal and cereal proteins. *International Journal of Food Science and Technology*, 39, pp. 183–189, ISSN 0950-5423

Gorinstein, S.; Drzewiecki, J.; Delgado, E.; Pawelzik, E.; Martinez-Ayala, A. L.; Medina, O. J.; Haruenkit, R. & Trakhtenberg, S. (2005). Relationship between dicotyledone-amaranth, quinoa, fagopyrum, soybean and monocots- sorghum and rice based on protein analyses and their used as substitution of each other. *European Food Research and Technology*, 221, pp. 69-77, ISSN 1438-2377

Hanelt, P. & Institute of Plant Genetics and Crop Plant Research (IPGCPR) (Eds.) (2001). Mansfeld's Encyclopedia of Agricultural and Horticultural Crops. 1-6, 3716 pp. ISBN: 978-3-540-41017-1, Berlin, Germany

'The International Plant Names Index' (IPNI) (2008). Available on the Internet http://www.ipni.org [accessed 22 August 2011].

Iqbal, S. M.; Ghafoor, A. & Ayub N. (2005). Relationship between SDS-PAGE Markers and Ascochyta Blight in Chickpea. *Pakistan Journal of Botany* 37 (1), pp. 87-96, ISSN 0556-3321

Jaramillo, S. & Baena M. (2002). Ex situ conservation of plant genetic resources: training module. International Plant Genetic Resources Institute, ISBN 978-92-9043-751-2, Cali, Colombia.

Jarošová, J.; Michalová, A.; Vavreinová, S. & Moudrý, J. (1997). Cultivation and utilization of amaranth. UZPI Praha, ISBN 80-7271-042-7. (in Czech)

Juan, R.; Pastor, J.; Alai, M. & Vioque, J. (2007). Electrophoretic characterization of *Amaranthus* L. seed proteins and its systematic implications. *Botanical Journal of the Linnean Society*, 155, pp. 57–63, ISSN 0024-4074

Jugran, A.; Bhatt, I. D. & Rawal, R. S. (2010). Characterisation of Agro-diversity by Seed Storage Protein Electrophoresis: Focus on Rice Germplasm from Uttarakhand Himalaya, India. *Rice Science*, 17 (2), pp. 122-128, ISSN1672-6308

Kaul, H. E.; Aufhammer, W.; Laible, B; Nalborczyk, E.; Pirog, S. & Wasiak, K. (1996). The suitability of amaranth genotypes for grain and fodder use in Central Europe. *Die Bodenkultur*, 47 (3), pp. 173 – 181, ISSN 0006-5471

Laemmli U. K. (1970). Cleavage of Structural Proteins during the Assembly of the Head of Bacteriophage T4. *Nature* 227 (5259), pp. 608-685, ISSN 0028-0836

Marcone, M. F. (2002). A non-DNA-based marker technique to differentiate between *Amaranthus pumilus* (a federally protected, threatened plant species) from among more common amaranth cultivars. *Journal of Food Composition and Analysis*, 15, pp. 115-119, ISSN 0889-1575

Michalova, A. (1999). *Amaranthus* L. Vyziva a potravinarstvi 54, pp. 13-14 (in Czech)

Moudry, J.; Pejcha, J. & Peterka, J. (1999). The effect of genotype and farming technology on the yield of amaranth *Amaranthus* sp. Collection of Scientific Papers, Faculty of Agriculture in ČeskeBudujovice, *Series for Crop Sciences* 16, pp. 93-98. (in Czech)

National Research Council (NRC) (1984). Amaranth: Modern Prospects for an Ancient Crop. National Academy Press, Washington, DC.80 pp.

Osborne T. B. (1907). The Proteins of the Wheat Kernel. Carnegie Institution of Washington, Washington: 119 pp.

Petr, J. ; J.; Michalik, I.; Tlaskalova, H.; Capouchova, I.; Famera, O.; Urminska, D.; Tuckova, L. & Knoblochova, H. (2003). Extention of the Spectra of Plant Products for the Diet in Coeliac Disease. *Czech Journal of Food Science*, 21 (2), pp. 59-70, ISSN 1212-1800

Pisarikova, B.; Zraly, Z.; Kracmar, S.; Trckova, M. & Herzig, I. (2005). Nutritive value of amaranth grain (*Amaranthus* L.) in the diets for broiler chickens. *Czech Journal of Animal Science*, 50, pp. 568–573, ISSN 1212-1819

T. & Roy, S. C. (2009). Genetic Diversity of Amaranthus Species from the Indo-Gangetic Plains Revealed by RAPD Analysis Leading to the Development of Ecotype-Specific SCAR Marker. *Journal of Heredity* 100 (3), pp. 338–347, ISSN 1465-7333

Rayburn, A. L.; McCloskey, R.; Tatum, T. C.; Bollero, G. A.; Jeschke, M. R.; & Tranel, P. J. (2005). Genome Size Analysis of Weedy Amaranthus Species. *Crop Science*, 45, pp. 2557–2562, ISSN 1435-0653

Segura-Nieto, M.; Barba de la Rosa, A. P. & Paredes-Lopez, O. (1994). Biochemistry of Amaranth Proteins. In: *Amaranth: Biology, Chemistry and Technology*. Paredes-Lopez O. (Ed.), pp. 76-106. CRC, ISBN 0-8493-5374-2, Boca Raton, USA

Sauer, J. D. (1950). The grain amaranths: a survey of their history and classification. *Annals of the Missouri Botanical Garden*, 37, pp. 561-619, ISSN0026-6493

Sauer, J. D. (1967). The grain amaranths and their relatives: a revised taxonomic and geographic survey. *Annals of MissouriBotanical Garden*, 54, pp. 103-137, ISSN0026-6493

Sangameswaran, B. & Jayakar, B. (2008). Anti-diabetic, anti-hyperlipidemic and spermatogenic effects of Amaranthus spinosus Linn. onstreptozotocin-induced diabetic rats. *Journal of natural medicines*, 62, pp. 79–82, ISSN 1340-3443

Smith, D. B. & Simpson P. A. (1983): Relationships of barley proteins soluble in sodium dodecyl sulphate to malting quality and varietal identification. *Journal of Cereal Science*. 1, 185-197, ISSN0733-5210

Srivastava, S. & Gupta, P. S. (2002). SDS and Native Page Protein Profile for Identification and Characterization of Elite Sugarcane Genotypes. *Sugar Technology* 4 (3), pp. 143-147, ISSN 0972-1525

Standley P. C. & Steyermark J. A. (1946): Flora of Guatemala. Series: Fieldiana: Botany 24 (4). 152 -157. Chicago, USA

Tironi, V. A. & Añón, M. C. (2010). Amaranth proteins as a source of antioxidant peptides: Effect of proteolysis. *Food Research International* 43, pp. 315–322, ISSN 0963-9969

Thanapornpoonpong, S.; Vearasilp, S.; Pawelzik, E. & Gorinstein, S. (2008). Influence of Various Nitrogen Applications on Protein and Amino Acid Profiles of Amaranth and Quinoa. *Journal of Agricultural and Food Chemistry*, 56, pp. 11464–11470, ISSN 1520-5118

Transue, D. K.; Fairbanks, D. J.; Robison, L. R. & Andersen W. R. (1994). Species Identification by RAPD Analysis of Grain Amaranth Genetic Resources. *Crop Science* 34: 1385 – 1389, ISSN: 1435-0653

Vasco-Mendez, N. L. & Paredes-Lopez, O. (1995). Antigenic homology between *Amaranth* glutelins and other storage proteins. *Journal of Food Biochemistry*, 18 (4), pp. 227-238, ISSN 0145-8884

Vaz, A. C.; Pinheiro, C.; Martins, J. M. N. & Ricardo, C. P. P. (2004). Cultivar discrimination of Portuguese *Lupinusalbus* by seed protein electrophoresis: the importance of considering "glutelins" and glycoproteins. *Field Crops Research*. 87, 23-34, ISSN 0378-4290

Xu, F. & Sun, M. (2001). Comparative Analysis of Phylogenetic Relationships of Grain Amaranths and Their Wild Relatives (*Amaranthus; Amaranthaceae*) Using Internal Transcribed Spacer, Amplified Fragment Length Polymorphism, and Double-Primer Fluorescent Intersimple Sequence Repeat Markers. *Molecular Phylogenetics and Evolution* 21 (3), pp. 372–387, ISSN1055-7903

Watson, S. (1877). Contributions to American Botany. VII. Descriptions of New Species of Plants, with Revisions of Lychnis, Erigonum, and Chorizanthe. *Proceeding of American Academy of Arts and Sciences*, 12, pp. 246-278

H.; Sun, M.; Yue, S.; Sun, H.; Cai, Y.; Huang, R.; Brenner D. & Corke, H. (2000). Field Evaluation of an *Amaranthus* Genetic Resource Collection in China. *Genetic Resources and Crop Evolution* 47 (1), pp. 43-53, ISSN 1573-5109

Zheleznov, A. V.; Solonenko, L. P. & Zheleznova, N. B. (1997). Seed proteins of the wild and the cultivated *Amaranthus* species. *Euphytica*, 97 (2), pp. 177-182, ISSN 0014-2336

9

Agronomic and Biotechnological Strategies for Breeding Cultivated Garlic in Mexico

Héctor Silos Espino et al.*

*Biotechnology Applied Laboratory, Instituto Tecnológico El Llano, Aguascalientes
(ITEL), km. 18 Carr. Aguascalientes-San Luís Potosí
México*

1. Introduction

Garlic is an apomyctic diploid species ($2n=2x=16$) with vegetative reproduction that belongs to the *Allium* genus (Alliaceae), which includes onion (*Allium cepa*), leek (*A. ampeloprasum*) and shallot (*A. ascalonicum*) (Mc-Collum, 1987; Figliuolo et al., 2001; Ipek et al., 2003; 2005). The importance of garlic was recognized by humans at bronze era about 5000 years ago, and since these early times, has been used as food, condiment and medicine by Asians and Mediterranean (Ipek et al., 2005). World production of garlic is ranked 14th among vegetables with a total of 14.5 million ton (Trejo, 2006). In Mexico, its consumption is about 400 g *per capita* (Chávez, 2008), and the national production is considered low as compared to other countries such as China (80 % world production), India, Korea and the rest of the world (20 %) (FAOSTAT, 2011). Still, Mexico has a place as exporter of garlic produced mainly from the states: Zacatecas, Guanajuato, Aguascalientes, Baja California, Puebla and Sonora (Trejo, 2006). The areas in Mexico during winter 2008, dedicated to garlic were 5,085 ha with a total yield of 49,968 ton (SIAP, 2011).

Among other problems, Mexican garlic has a limited spectrum of high yielding cultivars for different environments and, at the same time, have good market qualities. Keeping in mind that kind of problems, this chapter was mainly focused on agriculture and biotechnology research done at four institutions. The first two sections include morphological, physiological and cytogenetic characterizations of the most common cultivars and related germplasm; the third section describes some advances on garlic micropropagation. The last

* Flora San Juan Hernández[1], Olivio Hernández Hernández[1], Darío Silva Bautista[1], Alan Roy Macías Ávila[1], Francisco Nieto Muñoz[1], Luis L. Valera Montero[1], Silvia Flores Benítez[1], Luis Martín Macías Valdez[2], Tarsicio Corona Torres[3], Mario Leonel Quezada Parga[4] and Juan Florencio Gómez Leyva[5]
[1]*Biotechnology Applied Laboratory, Instituto Tecnológico El Llano, Aguascalientes (ITEL), km. 18 Carr. Aguascalientes-San Luís Potosí, México*
[2]*Instituto Nacional de Investigaciones Forestales, Agrícolas y Pecuarias. Campo Experimental. Pabellón de Arteaga, km 32.5 carr. Aguascalientes-Zacatecas, México*
[3]*Cytogenetic Laboratory "Maestra Czeslawa Prywer Linzbarka", Colegio de Postgraduados, km. 36.5, carr. México-Texcoco Montecillo, Texcoco, Estado de México, México*
[4]*Fundación Produce Aguascalientes A. C. Av. Universidad 604 Interior 6 y 7, Fracc. Unidad Ganadera, Aguascalientes Ags., México*
[5]*Molecular Biology Laboratory, Instituto Tecnológico de Tlajomulco, Tlajomulco de Zúñiga, Jalisco, Mexico*

section describes our strategy for obtaining garlic genotypes with higher yield capability and better bulb quality characteristics according to the market demand.

2. Origin and distribution

Garlic is native to India (Central Asia), where was considered a spice with mystical implications due to its medicinal attributes. Egyptian hieroglyphs and Roman texts refer to garlic as a source for health and strength required for physical work. During the Middle Ages was used to prevent cholera. Nowadays, it is known for its antiseptic, diuretic, vermifuge and vasodilator activities. It also stimulates bile and stomach secretions, and acts against atherosclerosis and thrombosis.

Spaniard conquerors carried with them garlic; first to Cuba and, later, to the rest of the American colonies. Early reports of garlic fields in Mexico appeared at the beginning of the twentieth century, and fifty years later, the central region of Mexico (Bajío) was the main area for garlic production. The time of harvest in that region made possible to start exporting surplus, since at that time of the year the world production is low.

Garlic species are widely distributed on boreal areas having temperate climates and mountainous areas from tropical regions. Most of the species diversity is found from Mediterranean countries to Central Asia. USA is considered as diversification center for *Allium* (Lagunes, 2009).

3. Plant description

Garlic (*Allium sativum* L.) is propagated asexually, but shows a high morphological diversity among cultivars. These cultivars have a range of adaptation to different environments (Paredes et al., 2008). Like onion, garlic plants have thin tape-shaped leaves about 30 cm long. Roots reach a 50 cm depth or little more. Heads or bulbs are white-skinned, divided into sections called cloves. Each head could have from 6 to 12 cloves, which are covered with a white or reddish papery layer or "skin". Bulbs are consumed fresh, totally or partially dried, and pickled. Although the bulb consumption is more common, tender shoots sometimes are a delicatessen for sophisticated cuisine. These shoots may be prepared like asparagus.

Each clove is capable to develop a new plant, since they have an apical shoot bud that can elongate even though if they are not sown. This shoot is apparent after three months after the harvest, depending on the genotype and conservation conditions. Flowers are white, and the stem of some species also produce small bulbils. These stems produce a strong odor from two compounds: alliin and diallylsulfide.

4. Mexican genotypes

The origin of present-day genotypes in Mexico was a group of cultivars: 'Perla', 'California', 'Chileno' and 'Taiwan', which were introduced from USA and China. A short description of each is included:

'Perla'. Late cultivar (240 d), with creamy-colored bulbs; 10 to 16 cloves per bulb covered with about seven outer layers. Plant height is 40-45 cm tall, having a pale-green open canopy. Experimental yields from this genotype usually range from 16 to 18 ton/ha. Physiological disorders are common, such as brush-like plant growth with excessive number of thinner leaves; the more severe this problem, the more the plant opens its canopy of leaves with reduced sheath. Bulbs of brush-like plants lose their covering layers, producing naked cloves. This disorder is high temperature-dependent, having the highest temperature influence on March and April; therefore, it varies in severity from year to year. Experimental observation indicates that some other factors alone or combined may be related to the induction of brush-like plants such as: early planting, excessive nitrogen fertilization, and planting density. This disorder worsens when these factors appear combined.

'California'. Late cultivar (260 d); recently introduced to Aguascalientes (Mexico). Bulbs are white, containing 18-26 cloves. Experimental yields range from 18 to 20 ton/ha. Plants are 50 cm tall on the average; leaves are pale-green with open canopy.

'Chileno'. Early cultivar (160 d), with a yield average of 7 ton/ha. Bulbs are purplish with 5-6 covering layers; containing 11-22 cloves (average = 19). Plants are about 50 cm tall; with semi-compact canopy and dark green leaves.

'Taiwan'. Early cultivar (170 d); its yield average is 7 ton/ha. Bulbs are purple in color, with 7-13 cloves (average = 9). Plant height reaches 50 cm on the average, with semi-compact canopy and dark green leaves.

5. Field performance of promissory genotypes

Besides the previously described genotypes, some more garlic accessions from the germplasm bank of INIFAP-CAEPAB (Fig. 1) were tested for their performance on the field (Aguascalientes, Mexico). These accessions have features suitable for breeding, as described below.

5.1 Bulb size and number of cloves per head

Higher values for bulb perimeter were found in 'California' (23.1 ± 1.8 cm) and 'Coreano' (20.4 ± 0.7 cm) varieties, as well as in the cultivars 'Perla' 'C-37-1/8'(21.1 ± 0.6 cm) and 'Perla' 'C-3-1/25' (20.5 ± 0.7 cm) (Table 1). Three varieties of white garlic, 'California' (112.3 ± 22.8 g), 'Perla' 'C-37-1/8' (84.3 ± 8.1 g) and 'Perla' 'C-3-1/25' (79.2 ± 7.8 g), as well as a marbled one, 'Coreano' (82.3 ± 8.0 g), showed the greatest bulb weight. Regarding the number of cloves per bulb, variety 'Español' produced 7.5 (±0.9), while cultivars 'Perla' 'C-3-1/25' and 'Perla C-37-1/8' had 10.9 (±1.3) and 11.9 (±1.9), respectively. Plants showing a smaller number of cloves per bulb appeared to have greater clove weights. 'Chino' and 'Coreano' varieties also showed a good clove weight performance. However, they are more susceptible to diseases and they require more time for bulb formation. Varieties with greater bulb weights appeared to be taller than those with smaller bulb weights (Table 1).

5.2 Days to harvest and yield

Varieties of garlic can be harvested at either 150 (early cycle), 180 (intermediate cycle) or 210 (late cycle) days after planting. Late cycle varieties showed greater bulb and clove weights [i.e., 'California' and 'Coreano' varieties, and 'Perla' cultivars (Table 1)]. Greatest bulb

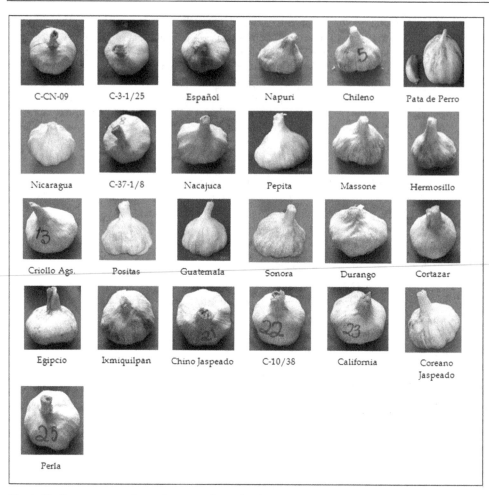

| C-CN-09 | C-3-1/25 | Español | Napuri | Chileno | Pata de Perro |

| Nicaragua | C-37-1/8 | Nacajuca | Pepita | Massone | Hermosillo |

| Criollo Ags. | Positas | Guatemala | Sonora | Durango | Cortazar |

| Egipcio | Ixmiquilpan | Chino Jaspeado | C-10/38 | California | Coreano Jaspeado |

Perla

Fig. 1. Garlic genotypes from the germplasm bank at Instituto Nacional de Investigaciones Forestales, Agrícolas y Pecuarias - Campo Experimental Pabellón Aguascalientes (INIFAP-CAEPAB).

weight varieties (i.e., 'California') showed more than 75% greater bulb weight than lowest bulb weight varieties (i.e., 'Pata de Perro'). 'Perla' cultivars ('C-3-1/25' and 'C-37-1/8') had a better tolerance to environmental conditions (Fig. 1), their bulbs had fewer cloves (10-12) (Table 1), and their bulb and clove weights were favorably compared with those of commercial varieties (i.e., 'Chino' and 'Coreano': Table 1). 'Sonora', 'Positas', 'Hermosillo', 'Español', 'Pepita', 'Massone', 'Nicaragua', 'Nacajuca' and 'Chileno' cultivars showed very similar patterns in morphological characteristics and yields (Table 1). Correlation between bulb length and clove weight against bulb weight showed an $R^2 = 0.9668$, while clove weight against bulb weight had an $R^2 = 0.635$ (both highly significant, $p < 0.01$); in agreement with results of Heredia & Delgadillo (2002), these varieties which showed greater bulb weights also showed greater clove weights.

Genotype	Bulb length (cm)		Bulb weight (g)			Clove number/bulb		Plant length (cm)		Clove weight (g)	
California 20/1	23.1	± 1.8	112.3	±	22.8	17.4	± 4.6	71.1	± 2.9	8.1	± 2.7
Coreano	20.4	± 0.7	82.3	±	8.0	12.9	± 3.5	67.6	± 4.9	9.9	± 2.0
Perla C-3 - 1/25	20.5	± 0.7	79.2	±	7.8	10.9	± 1.3	74.3	± 3.4	12.4	± 1.1
Perla C-37 - 1/8	21.1	± 0.6	84.3	±	8.1	11.9	± 1.9	73.9	± 3.4	12.8	± 3.7
Chino	18.7	± 0.9	62.5	±	6.7	12.2	± 1.4	49.2	± 3.8	7.2	± 1.1
Ixmiquilpan	18.7	± 1.6	60.8	±	10.4	21.8	± 1.3	72.9	± 2.4	3.7	± 1.1
Durango	19.9	± 1.0	71.9	±	7.7	19.9	± 3.3	75.7	± 3.6	4.8	± 1.0
Criollo Ags.	17.4	± 1.1	49.7	±	8.3	14.5	± 1.4	63.9	± 2.9	5.3	± 2.6
Cortazar	18.4	± 0.7	57.4	±	5.8	20.7	± 2.3	64.7	± 2.9	3.3	± 0.5
Sonora	14.9	± 0.7	36.4	±	5.2	16.9	± 3.5	47.7	± 5.4	2.5	± 0.8
Guatemala	14.8	± 1.3	33.8	±	8.8	14.8	± 2.5	62.0	± 5.4	3.0	± 0.7
Positas	14.3	± 0.7	34.6	±	4.9	14.7	± 3.5	51.5	± 3.2	3.7	± 0.9
Hermosillo	13.9	± 1.3	31.9	±	6.7	12.6	± 3.7	60.1	± 6.6	2.8	± 0.8
Español	13.6	± 1.4	24.6	±	5.7	7.5	± 0.9	59.9	± 2.4	4.2	± 1.0
Pepita	14.0	± 0.6	32.9	±	4.2	20.0	± 4.0	48.3	± 5.6	2.5	± 0.6
Massone	13.2	± 0.6	30.2	±	3.6	15.0	± 2.1	48.1	± 4.6	2.9	± 0.8
Nicaragua	14.1	± 0.9	29.6	±	6.6	13.0	± 2.7	51.5	± 3.2	2.7	± 0.8
Nacajuca	14.4	± 1.1	31.9	±	6.7	18.4	± 3.7	54.7	± 7.0	2.9	± 0.8
Chileno	14.2	± 1.2	32.4	±	5.8	17.6	± 5.7	49.2	± 3.8	2.8	± 0.5
Napuri	13.5	± 1.3	31.3	±	7.1	14.8	± 4.9	47.0	± 2.9	3.0	± 0.7
Pata de Perro	13.3	± 0.7	27.7	±	4.4	8.4	± 1.2	55.0	± 3.6	3.4	± 1.3

Table 1. Size and weight of garlic varieties cultivated in Aguascalientes (Central-North Region of Mexico). Data are presented according to standard descriptors for garlic (IPGRI, 2001). Each value shows the mean ± SE.

5.3 Postharvest photosynthesis and respiratory activity of stored cloves

In order to understand some physiological events of stored garlic, analyses of six genotypes under storage were focused on the respiratory process and photosynthesis. To accomplish that goal, photosynthetic activity of stored cloves during 0, 30, 60 and 90 d were measured on three cloves selected at random from the container of each genotype. Measurements included: evaporation rate mM/s/m2/s (E), stomatal conductance mM/m2/s (G), net photosynthesis assimilation $\mu M/m2/s$ (A) and CO_2 internal concentration ppm (CI). From these measurements, it was found that for some genotypes like 'C-CN-9/2' evapotranspiration was the highest at 30 d, as opposed to 'Criollo Aguascalientes' and 'Chino Jaspeado' with the lowest value at 90 d. Stomatal conductance was high 'C-CN-9/25', mainly after 90 d. Most genotypes showed negative photosynthesis rate, and internal CO_2 showed no clear tendency within genotypes. Weight remained stable during the first 60 d, but after that period, it decreased about 1/3 of the initial values. The bulbs behavior at the final of postharvest period is show in the Fig. 2.

Fig. 2. View of stored garlic at 90 d. Sprouted heads are from short-shelf life cultivars. Heads with low number of sprouting bulbs are from 'Perla' (bottom left corner).

6. Isolation and culture of garlic protoplasts

Garlic genotypes derived from 'Perla' (C-37-1/8, C-3-1/25 and C-CN-95/2), 'Chino', 'Coreano' and 'Criollo Aguascalientes' were used for protoplast preparation. Cloves of these genotypes were peeled and disinfected (briefly: cloves were soaked in 70% ethanol 1 min and 20% commercial bleach 20 min, after three rinses with sterile distilled water the cloves were soaked again in 70% ethanol 1 min and rinsed again with sterile distilled water). Surface-sterilized cloves were cut into pieces 0.5 to 1.0 cm long, and were inoculated aseptically on MS (Murashige & Skoog, 1962) medium pH 5.8 containing 0.15 mg L^{-1} 2,4-D (Dichlorophenoxy acetic acid), 5.0 mg L^{-1} BA (Benzyl adenine), 50 g L^{-1} glucose and 3.5 g L^{-1} Phytagel. Incubation conditions were 28 ±2 °C, with a light intensity of 59 μEm^2s.

Protoplasts were isolated from leaves and callus using a modification of the method from Spangenberg (1997). Briefly: 1 g of tissue was placed in petri dish together with 5 ml of enzyme mixture (2% Onozuka R-10, 1% macerozyme, 0.5% pectinase, 0.5% mannitol and 0.9% $CaCl_2$); dishes were incubated on orbital shaker at 5 rpm, about 4-6 h (Novák et al., 1983). Callus tissues were best for protoplast isolation within the range 10^6-10^8 counts/ml (Fig. 3). Protoplasts were isolated from the debris and inoculated on semi-solid MS medium. Unfortunately, whole plant regeneration remained elusive.

Genotype	days	Clove weight (g)		Evaporation rate (mM/s/m²/s)		Stomatal conductance (mM/m²/s)		Net photosynthesis assimilation (μM/m²/s)		CI (ppm)	
		Mean	SD	Mean	SD	Mean	SD	Mean	SD	Mean	SD
'C-CN-9/2'	0	65.30	8.21	69	21	45	15.5	-2.1	0.5	267	49.5
	30	64.6	9.7	0.3	0	14	1	-1.6	2.9	327	16.9
	60	64.0	8.3	0.5	0.7	40	59	-5.6	7.1	1332	1938.2
	90	19.4	30.4	7.6	0.8	6618.3	2929.3	-1.6	0.4	192.7	6
'C-3-1/25'	0	68.50	7.27	0.5	0	23.7	4	-2.5	1.1	306.3	56.9
	30	61.1	4.4	0.3	0.1	13.3	3.8	0.3	3.7	247.7	223.1
	60	67.2	7.0	0.4	0.1	28.3	3.2	-3.8	1.3	372.7	66
	90	15.6	31.8	1.6	1.3	122.7	138.5	-3.5	1.3	313.7	120.5
'C-37-1/8'	0	66.35	17.22	1.3	0.6	75.7	27.1	-1.6	0.6	288.7	15.2
	30	63.2	4.3	0.2	0.2	8.7	7.6	-14.4	17.2	3849	5328
	60	65.1	16.4	41	23.4	27.7	1.5	-4.7	3.4	356	289.2
	90	21.8	28.7	4.6	3.6	1064.7	1210	7.5	15.5	295	119.8
'Chino Jaspeado'	0	89.93	12.51	0.3	0.1	20.7	10.8	-3	0.5	492.7	309.5
	30	92.3	15.7	0.6	0.3	30.3	13.5	-2.9	1.3	346.3	46.2
	60	88.5	12.0	1.1	0.2	78.3	14.6	4.4	7.9	156.3	50.3
	90	28.6	42.2	0.1	0.1	6.3	5.5	-1	2.7	0	0
'Coreano Jaspeado'	0	94.38	11.73	0.2	0.1	18.3	10.5	-1	2.1	333.3	140.8
	30	92.2	8.7	0.4	0.2	16.7	7.1	-0.2	2.6	233.7	212.7
	60	93.1	11.4	2.1	1.1	152.3	96	-8.3	2.6	311.3	131.5
	90	25.1	45.0	4.1	5.4	3371	5740	-3.8	6.1	315	103.6
'Criollo Aguascalientes'	0	48.83	7.77	0.4	0.1	23.7	5.7	-1.5	0.3	309.3	68.2
	30	51.0	7.2	0.2	0.2	7.7	6.7	-2.3	2.5	3678.7	5477.2
	60	48.0	7.5	0.4	0.2	26	12.5	-8.2	9.3	471.7	40.1
	90	15.9	23.0	0.1	0	5.3	2.9	-9.6	4.1	1899.3	1388.7

Table 2. Photosynthetic and respiration rate of cloves from six garlic genotypes after 90 d storage.

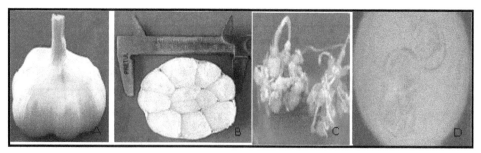

Fig. 3. Protoplast isolation from P-C-3 1/25 genotype: A,B) Bulb ('Perla' type), C) *In vitro* callus culture and D) Protoplasts at 40x magnification.

7. Karyotyping Mexican garlic genotypes

Karyotypes of C-CN-95/2, C-37 1/8, C-3 1/25 (all these 'Perla' type genotype), 'Chino', 'Coreano' y 'Criollo' were obtained from root tips. Cloves of these genotypes were placed inside petri dishes containing wet cotton wool in order to induce roots 1-2 cm long. Roots were removed and soaked with 0.05% w/v colchicine and placed in the darkness for 3:30 h at 25°C. These roots were fixed with Farmer´s solution (ethanol and glacial acetic acid, 3:1 v/v); then, they were hydrolyzed with 1N HCl at 60 °C for 10 min. Feulgen stain was applied to fixed roots before maceration with an enzymatic solution (2% pectinase, 5% celulase and citrate buffer pH 4.5) for 30 to 60 min. Roots tips were placed on microscope glass slides with a drop of 2% propionic orcein, and sandwiched with a cover glass. The slides were heated for few seconds with an alcohol burner with a very soft press so that single cells could be freed from the tissue. Then the cover glasses were gently tapped with a pencil in order to squeeze single cells for releasing and spreading the chromosomes. Observation of microphotography 100x25" allowed the following counts and measurements: chromosome number, short arms (p), long arms (q) total length, relative size of the chromosome and arm relationship (García, 1990). All of these observations were useful to classify each garlic genotype according to karyotype nomenclature and formule from Levan et al., (1964).

According to the observations, all of the genotypes tested have a chromosome number $2n=8x=16$ (Fig. 4) in agreement with other reports (Battaglia, 1963; Verma & Mital, 1978; Koul et al., 1979). The karyotype characteristics found for centromer position were as follows: 'P-CN-95/2' (1M+4m+3sm), 'P-37-1/8' (1M+6m+1st), 'P-C-3-1/25' (7m+1sm), 'Chino' (1M+3m+3sm+1st), 'Coreano' (6m+2sm) and 'Criollo' (5m+3sm); Code: M or m=metacentric, sm= submetacentric and st=subtelocentric (Table 3). The nuclear content (2C value), which is considered one of the highest among the cultivated plants (Ipek et al., 2005), is 32.7 pg. Additionally, garlic has a high karyotype variability that may be attributed to repetitive ADN (Kirk et al., 1970).

8. Genetic profile of Mexican garlic

Genetic markers are efficient tools for genetic analysis of populations and individuals. According to this concept, molecular characterization of garlic around the world has been performed either through RAPDs (Bradley & Collins, 1996; Eom & Lee, 1999; Shasany et al.,

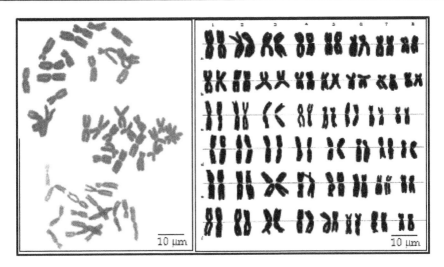

Fig. 4. Chromosomes from somatic cells (n=8x=16) from garlic genotypes: 'C-CN-95/2', 'C-37- 1/8', C-3 1/25, 'Chino', 'Coreano' and 'Criollo'.

2000; Peiwen et al., 2001; Ipek et al., 2003; Paredes et al., 2008; Pardo et al., 2009) or through AFLP (Rosales & Molina, 2007). Our version of this kind of analysis with 20 Mexican genotypes was as follows: Twenty garlic genotypes were subjected to RAPDs in order to construct a distance tree using clustering with the Unweighted Pair Group Method with Erithmetic Mean (UPGMA). DNA was extracted according to Doyle and Doyle (1990); DNA samples were run on agarose gel 0.8% and DNA concentration was measured with a spectrometer (model GBC Cintra 10e UV-visible). RAPD reactions were performed in a 25 ml volume, consisting of 10x buffer solution [10 mM Tris-HCl buffer (pH 8.0), 50 mM KCl₂], 2.5 mM MgCl₂, 2.5 units of Taq DNA polymerase (Promega), 100 µM dNTP, 50 ng genomic DNA and 0.4 µM OPB series (OPB-8, OPB-9, OPB-10, OPB-11, OPB-15 and OPB 17) primer (Operon Technologies, Alameda, CA, USA). A total of 20 µl of mineral oil was placed over the reaction mixture. Amplifications were carried out in a DNA thermocycler (Model FPR0G02Y Techne Progene, England), under the following conditions: an initial denaturalization step of 2 min at 94 °C, followed by 35 cycles of 1 min at 94 °C, 1 min at 35 °C, and 2 min at 72 °C, with a final extension step of 7 min at 72 °C. Amplification products were analyzed by electrophoresis in a 1.2% agarose gel. It was run at 100 V for 4 h, and detected by staining the gel with ethidium bromide (10 ng/100 ml of agarose solution in TBE). All visible and unambiguous fragments amplified by the chosen primers were entered under the heading of total visible fragments. Fragment data were entered on a spreadsheet to form a binary matrix, where (1) represented fragment presence and (0) fragment absence for each fragment accession combination. Cluster analysis was conducted by converting the data matrix into a similarity matrix using a simple matching coefficient. This coefficient was calculated by dividing the number of matches (0-0 and 1-1) by the total number of comparisons (Nei & Li, 1979). A cluster analysis was then conducted using the unweighted pair group method, with arithmetical averages (UPGMA) process using the S–Professional Plus 2000 program. The results obtained were compared with others studies realized by different authors, and were discusses as following: Six decamer OPB primers showing

Chrom. no.	LBL	LBC	LT	LR	r	N	Chrom. no.	LBL	LBC	LT	LR	r	N
'P-CN-95/2'							Chino'						
1	6.75	2.5	9.25	0.12	2.7	sm	1	6.75	4	10.75	0.13	1.7	sm
2	6.5	2.5	9	0.11	2.6	sm	2	6.5	6.25	12.75	0.16	1.04	m
3	6	6	12	0.15	1	M	3	6.5	2	8.5	0.10	3.25	st
4	6	5.5	11.5	0.15	1.09	m	4	6	6	12	0.15	1	M
5	6	5	11	0.14	1.2	m	5	6	5.5	11.5	0.14	1.09	m
6	5.5	4	9.5	0.12	1.3	m	6	5.5	4.75	10.25	0.13	1.2	m
7	5	4.5	9.5	0.12	1.1	m	7	5.5	3.5	9	0.11	1.6	sm
8	4.5	2.5	7	0.09	1.8	sm	8	4.5	2.5	7	0.08	1.8	sm
Formula: 1M+4m+3sm							Formula: 1M+3m+3sm+1st						
'P-37-1/8'							'Coreano'						
1	6.75	2	8.75	0.14	3.4	st	1	7	5.5	12.5	0.15	1.2	m
2	5	4.5	9.5	0.15	1.1	m	2	7.5	4	11.5	0.14	1.8	sm
3	5	4	9	0.14	1.2	m	3	7	2.5	9.5	0.11	2.8	sm
4	4	4	8	0.13	1	M	4	6.5	6	12.5	0.15	1.08	m
5	4.5	2.75	7.25	0.12	1.6	m	5	6.5	5	11.5	0.14	1.3	m
6	4	3.25	7.25	0.12	1.6	m	6	5	4.5	9.5	0.11	1.1	m
7	4	3	7	0.11	1.3	m	7	5	4.5	9.5	0.11	1.1	m
8	3.5	2.25	5.75	0.09	1.5	m	8	4.5	3	7.5	0.09	1.5	m
Formula: 1M+6m+1st							Formula: 6m+2sm						
'P-C-3-1/25'							'Criollo'						
1	6.75	4	10.7	0.14	1.7	m	1	7	5.5	12.5	0.15	1.2	m
2	6.5	5.5	12	0.15	1.2	m	2	7.5	4	11.5	0.13	1.8	sm
3	6	5	11	0.14	1.2	m	3	6.5	6	12.5	0.15	1.0	m
4	5.75	5.5	11.25	0.15	1.04	m	4	6.5	6	12.5	0.15	1.0	m
5	5.75	2.75	8.5	0.11	2.0	sm	5	6.5	4.5	11	0.13	1.4	m
6	5	4	9	0.12	1.3	m	6	6.5	2.5	9	0.10	2.6	sm
7	5	3.5	8.5	0.11	1.4	m	7	5.5	3.25	8.75	0.10	1.7	sm
8	4	2.5	6.5	0.08	1.6	m	8	4.5	3.5	8	0.09	1.3	m
Formula: 7m+1sm							Formula: 5m+3sm						

Table 3. Chromosome morphological description from Mexican garlic genotypes. LBL=Long arm, LBC=Short arm, LT=Total length, LR=Relative length, r=Arm relationship and N=Centromere nomenclature.

distinct polymorphic fingerprint were selected to reveal the genetic variation among the garlic samples. In almost all varieties, it was possible to identify around 10 bands. A dendrogram was generated from the binary matrix of measured data (Fig. 5), and two groups were identified. The first group was formed by eight varieties ('Durango', 'Nicaragua', 'Cortazar', 'Hermosillo', 'Massone', 'Pepita', 'Sonora' and 'Napuri') that are characterized by a lower production (smaller clove weight and/or greater number of cloves: Fig. 6), required more days to dormancy (6 months) and need fewer days (150) to harvest (data not shown). The second group was constituted by white, colored and marbled garlic ('Coreano', 'Positas', and 'Perla' cultivars, 'Criollo Aguascalientes', 'Español', 'Chileno', 'Ixmiquilpan', 'California', 'Chino', 'Pata de Perro' and 'Guatemala'). These are characterized by better bulb and clove weights, lower clove numbers/bulb, fewer days to dormancy (5-6 months), and between 180-210 days to harvest. In general, garlic varieties were clustered according to yield level, clove and bulb weights, number of cloves/bulb and dormancy period (Table 1). These results agree with those of García et al., (2003) using the AFLP technique. The most productive variety ('California') has the inconvenient of having a larger number of cloves/bulb and requires lower temperatures to achieve complete bulb formation. Dissimilarity among the two groups was 0.33. The lowest dissimilarity (0.0) corresponded to the most related varieties ('Sonora'-'Napuri', 'Criollo Aguascalientes'-'Español', 'California'-'Chino' and 'Pata de Perro'-'Guatemala'). The highest dissimilarity (0.70) was between the variety 'Cortazar' and the varieties 'Ixmiquilpan', 'Pata de Perro', and 'Guatemala'. (Choi et al., 2003) reported a dissimilarity of 0.4 between two large groups from a total of 75 garlic varieties using the "M" or affinity coefficient. Using the Jaccard coefficient, Ipek et al., (2003) obtained a lowest dissimilarity of 0.0 and a highest dissimilarity of 0.75 between two large groups of garlic. These results are similar to those of Al-Zaim et

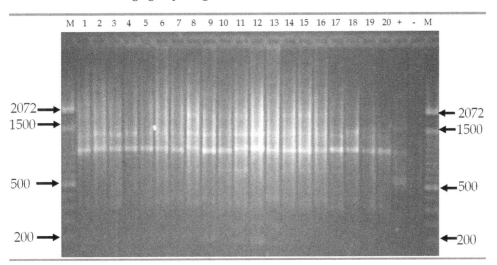

Fig. 5. RAPD from Mexican genotypes. Lanes: 1: C-3-1/25, 2: C-37-1/8, 3: 'Coreano', 4: 'California', 5: 'Chino', 6: 'Criollo Aguascalientes', 7: 'Español', 8: 'Cortazar', 9: 'Positas', 10: 'Pepita', 11: 'Massone', 11: 'Durango', 13: 'Chileno', 14: 'Hermosillo', 15: 'Sonora', 16: 'Nápuri', 17: 'Nicaragua', 18: 'Ixmiquilpan', 19: 'Pata de perro', 20: 'Guatemala', (+): positive control, (-): negative control and M: Molecular weight marker.

al., (1997). Evaluating diversity and genetic relationships among the progenitor *A. longicuspis* and 27 garlic varieties collected from different regions of the world, these authors found a dissimilarity of 0 between two samples, and a highest dissimilarity of 0.82 between two large groups. Through the RAPD technique used in this work, the two 'Perla' cultivars were grouped with the best production varieties, where the two selections presented a dissimilarity of 0.1. However, the 'Perla' C-3-1/25 cultivar showed a band of 2100 bp, and could thus be identified as a possible molecular marker. Our results allowed to identify highly related garlic varieties ('Sonora'-'Napuri', 'Criollo Aguascalientes'-'Español', 'California'-'Chileno' and 'Pata de Perro'-'Guatemala'), and separate them from varieties that are characterized by a lower yield (i.e., 'Pata de Perro' and 'Napuri'), and from mixed garlic that has been generated from introduced commercial varieties ('Criollo

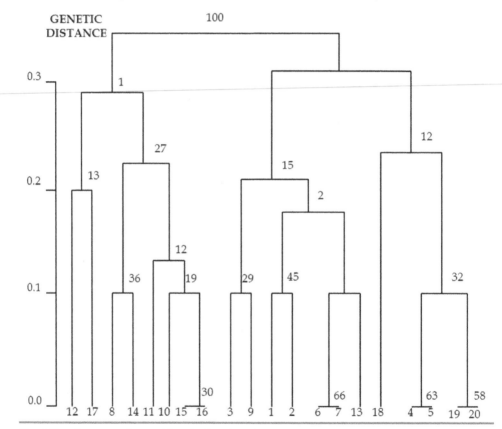

Fig. 6. Dendrogram obtained by the RAPD technique and general description according to physiological, morphological and genetic characteristics in garlic (*Allium sativum* L.) varieties cultivated in the Central Region of Mexico. Arms or branches of the dendrogram: 1. 'Perla' C-3-1/25, 2. 'Perla' C-37-1/8, 3. 'Coreano', 4. 'California', 5. 'Chino', 6. 'Criollo Aguascalientes', 7. 'Español', 8. 'Cortazar', 9. 'Positas', 10. 'Pepita', 11. 'Massone', 12. 'Durango', 13. 'Chileno', 14. 'Hermosillo', 15. 'Sonora', 16. 'Napuri', 17. 'Nicaragua', 18. 'Ixmiquilpan', 19. 'Pata de Perro' 20. 'Guatemala'.

Aguascalientes'); these have lost their potential yield because they have not been subjected to an appropriate selection process. Because we do not know the source of most garlic varieties and cultivars used in this work, we cannot establish relationships among their origins.

9. *In vitro* propagation

Most Mexican garlic cultivars are somewhat susceptible to pests and diseases. Furthermore, garlic is a seedless plant that could carry diseases to the next generation through vegetative propagation. "Seed cloves" per hectare range from 200,000 to 250,000 for typical plant density; therefore, totally clean vegetative material may be too difficult to be generated. The solution for this kind of problems is considered under biotechnological view such as *in vitro* culture so that we can obtain homogeneous healthy plants. *In vitro* bulbils after four years may produce, depending on the cultivar, from 1,400 to 10,700 bulbils ready to be used as "seed" (Burba, 1993). A recent report mentioned plants produced from cloves cultivated *in vitro* with 1 mg L^{-1} TDZ-1 (Thidiazuron), 1 mg L^{-1} GA$_3$ (Gibberelic acid) and coconut milk (Lagunes, 2009). *In vitro* garlic plants were also obtained on MS supplemented with 2.0 mg L^{-1} 2iP (2-isopentenyl adenine), 0.1 mg L^{-1} NAA (Naphthalene acetic acid) and 30 g L^{-1} sucrose (Mujica et al., 2008). Basal plate from cloves has the highest callus production as compared to leaves, stem segments, pedicels and aerial bulbils (Rabinowitch & Brewster, 1990). Another report mentioned that callus formation from 'Rojo de Cuenca' was the best on media having BA and NAA; furthermore, high BA concentration promoted adventitious shoot formation, but did not show influence on callus formation (Barandiaran et al., 1999). Callus was also obtained from leaves exposed to 0.3 a 0.5 mg L^{-1} 2,4-D (Fereol et al., 2002).

Our own results showed that explants about 5 mm^2 of clove sections were enough to regenerate *in vitro* whole plants from 'C-3-1/8' and 'C-37-1/25', 'Chino' and 'Coreano' garlic genotypes. The first step began with explants for root production, that were placed into MS medium supplemented with combinations of 0.15 mg L^{-1} 2,4-D, 5 mg L^{-1} adenine, 1.4 mg L^{-1} 4-amino-3,5,6-trichloropicolinic acid (Pichloram) and 1 mg L^{-1} 6-(γ,γ-dimetilamino) purine (2iP) (Table 4). MS medium also contained 30-50 g sucrose and 3.5 g L^{-1} Phytagel. Cloves were soaked in 70% ethanol 1min; after that, were transferred to 20% commercial bleach for 20 min. Then, cloves were rinsed thrice with dH$_2$O, placed back to 70% ethanol 1 min and rinsed again. Incubation conditions were: 28 ±2 °C and 16/8 h photoperiod. Adventitious roots were used to produce protocorms and protocorm-like bodies. These protocorms were cultured for 30 weeks on MS supplemented with 1 mg L^{-1} IAA and 5 mg L^{-1} adenine.

9.1 Protocorm formation

Protocorms 0.5-1 cm merged from root tips after 8 weeks on MS supplemented with 1.5 mg L^{-1} 2,4-D and 5 mg L^{-1} adenine (Table 4). Then they were placed into basal MS during 3 weeks, before protocorms were inoculated into four regeneration media at 18 ±4 °C and 16/8 h photoperiod during six weeks (Capote et al., 2000; Robledo-Paz et al., 2000; Quintana-Sierra et al., 2005). Protocorm and protocorm-like structures were both light and dark-green colored, compact and easily detachable (Fig. 7a, b and c), somehow similar to organogenic callus from *A. cepa* (Van der Valk et al., 1992; Zheng et al., 1998). 'Chino' and 'Coreano' protocorms were even more easily detached, dark-green colored as compared to 'Perla' derived genotypes; Novák et al., (1986) also found differential pigmentation among genotypes. Media supplemented with Pi (1.4 Mg/L^{-1}) and 2iP (1 Mg/L^{-1}) only promoted long roots.

9.2 Microbulbil formation

The treatment supplemented with 2,4-D showed the highest number of protocorm formation (Table 4); when the treatment included IAA and adenine, 'Chino' and 'Coreano' doubled to genotypes 'C-3-1/8' 'C-37-1/25'. Other treatments induced root formation. This different varietal response was found by Capote et al., (2000) and Quintana-Sierra et al., (2005) for *Allium cepa* and would be related to differences in sensibility to growth regulators (Fehér et al., 2003). Microbulbils placed, during three weeks, on basal MS (no regulators) increased their size (Fig. 7d), but after 30 weeks they developed into bulbils 1 cm diameter (Fig. 7e and 7f) and finally grew into whole plants.

Media	Growth regulators (mg/$^{-1}$)		Protocorm	Number of observed structures			
(%)	Auxins	Cytokinins	induction (%)	'C-3-1/8'	C-37-1/25'	Chino'	'Coreano'
Protocorms							
MS 100	2,4-D (1.5)	Adenine (5)	90	4	4	8	8
MS 100	Pi (1.4)	2iP (1)	0	----	----	----	----
Microbulbils							
MS 100	0	0	20	----	----	----	----
MS 100	IAA (1)	Adenine (5)	80	2-4	2-4	4-8	4-8
MS 100	NAA (1)	Kin (2)	0	----	----	----	----
MS 100	IAA (1)	Kin (2)	0	----	----	----	----
MS 100	Pi (1.4)	2ip (1)	0	----	----	----	----

Table 4. Effect of growth regulators on protocorm and microbulbil formation of four garlic 12 genotypes. IAA: Indole acetic acid, NAA: Naphthalene acetic acid, Kin: Kinetin, 2,4-D: 2,4-13 Dichlorophenoxyacetic acid, Pic: Pichloram, 2iP: 2-isopentenyl adenine.

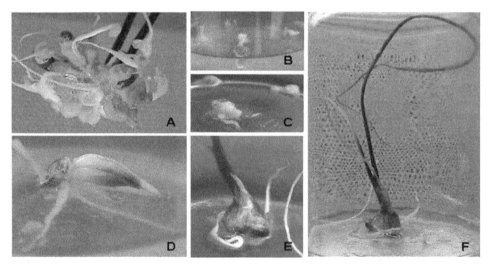

Fig. 7. *In vitro* regeneration of garlic plants (*A. sativum* L.). A) Protocorms regenerated from root tips, B) 'Chino' protocorms, C) 'Perla' protocorms, D) Microbulbil, E) and F) 'Chino' whole plant.

10. Individual selection for breeding Mexican garlic cultivars

Garlic in some cases may produce inflorescences but infertile seed; hence, crosses are not possible. Sometimes, bulbs o bulbils develop onto inflorescences (Brewster, 1994). Individual selection on best plants (yield or quality) has been used for breeding (Heredia and Heredia, 2000; González, 2006; Con, 1997). The CAEPAB group worked with individual selection from 'Perla' clones and 'Chileno' having heads with fewer cloves than the average for the original cultivar taking into account also: head size, vigor (hardiness) and plant healthiness. This initial work led to obtain two garlic cultivars: 'San Marqueño' from line 'C-37-1/8' (Macías et al., 2009) and 'Diamante' from line 'CAL-RN-11-1-2-4' derived from an Aguascalientes-Zacatecas collection (Macías & Maciel, 2003). In brief methodology for garlic breeding:

1. Bulb collection of promissory plants from the fields of outstanding growers of Aguascalientes and Zacatecas (May 1999).
2. Bulbs were planted in the experimental fields at CAEPAB, in order to check all of the collected material under the same growing conditions.
3. Evaluation and selection of garlic plants during 6-8 years (Table 6).
4. Storage of best clones at the germplasm bank (CAEPAB).
5. "Seed" production, enough to be transferred to growers for commercial validation.
6. After validation, best genotypes, having consistent yield results through time, are released to farmers.

Clones obtained through this kind of breeding are grown by farmers from Aguascalientes and Zacatecas (Macías et al., 2009). Nowadays, 'San Marqueño' and 'Diamante' garlic are demanded in Europe because of their high quality that makes them suitable to be exported (Fig. 8). These clones have their optimal conditions at 2000 meters over the sea level, on loamy soils, well drained, without salinity or pedregosity. Lab test are encouraged to check for soil pathogens that may reduce yield.

Evaluation	San Marqueño		Diamante	
Year	Yield (kg/ha)	SD	Yield (kg/ha)	SD
1	16,072	1,975	18,285	3,470
2	19,296	3,115	21,785	3,828
3	16,148	2,778	22,254	4,144
4	15,689	2,537	22,323	3,188
5	17,735	3,510	20,321	3,999
6	17,064	2,571	20,731	3,304
7			16,405	3,383
8			24,043	4,336

Table 6. Evaluation of 'San Marqueño' and 'Diamante' garlic cultivars.

Fig. 8. Garlic cultivars from the Mexican breeding program (CAEPAB): A) 'San Marqueño' ('Perla type'), B) 'Diamante' ('California' type) and C) Packaging box for exportation.

11. Conclusion

Horticulturists around the world look for answers from experimental stations to problems such as low yields, pests, diseases and quality defects. Similarly, garlic growers from Central Mexico have been in contact with institutions such as CAEPAB, PRODUCE-Ags and ITEL, in order to agree on agronomic and biotechnological research that may be applied to their fields. Original garlic genotypes from these growers and some other introduced to Mexico were the source for new cultivars and promissory genotypes developed by CAEPAB. Some of these that were analyzed showed a good correlation between bulb sizes and clove weight against bulb weight. It was also found that Mexican genotypes have a wide variety of on clove size and number that reflects a good genetic pool for breeding through individual selection for this seedless plant. Some other characteristics are qualitative that may have positive impact for worldwide market demands. For example, it was found for late cultivars the longest storage life.

Garlic biotechnology was also directed to characterize of Mexican cultivars. Molecular and cytogenetic characterizations for these cultivars may help to identify and to register, unambiguously, cultivated varieties. Molecular analysis such as AFLP's or RAPDs is required in order to establish genetic uniqueness or relatedness. So far, only RAPDs have been performed but this kind of work is not concluded yet. Another biotechnological application is to produce *in vitro* pathogen-free vegetative material for massive propagation that may be released to growers. Protocorm and protocorm-like bodies were produced *in vitro* before they grew into bulbils and whole plant. Nevertheless, massive propagation of garlic has not been achieved as desired. Therefore, garlic biotechnology is going at slow pace. Finally, breeding through individual selection allowed releasing cultivars appropriate for national and international demands.

12. Acknowledgements

The authors would like to thank Consejo de Ajos de Aguascalientes A. C. (Refugio Lucio Castañeda, Gerardo Vargas Prieto, Otilio Torres Robles, and Narváes brothers) for the facilities provided (in their cultivation áreas) to carried out some research activities here showed.

13. Reference

Al-Zahim, M., Newbury, H.J. & Ford-Lloyd, B.V. (1997). Classification of genetic variation in garlic (*Allium sativum* L.) reveled by RAPD. *The American Society for Horticultural Science.*Vol. 32, No. 6, pp. 1102-1104, ISSN: 0018-5345.

Al-Zahim, M.A., Ford-Lloyd, B.V. & Newbury, H. J. (1999). Detection of somaclonal variation in garlic (*Allium sativum* L.) using RAPD and cytological analysis. *Plant Cell Reports,* Vol. 18 , pp. 473-477.

Barandiaran, X., Di Pietro, A. & Martin, J. (1998). Biolistic transfer and expression of a *uid* reporter gene in different tissue of *Allium sativum* L. *Plant Cell Reports,* Vol. 17, No. 6, pp. 737-741, ISSN: 0721-7714.

Battaglia, E. (1963). Mutazione cromosomica e cariotipo fondamental in *Allium sativum* L. Caryologia. pp.16: 1-46.

Bradley, K.F., Rieger, M.A. & Collins, G.G. (1996). Classification of Australian garlic cultivars by DNA fingerprinting. *Australian Journal of Experimental Agriculture*, Vol. 36, No. 5, (1996), pp. (613-618), ISSN: 0816-1089.

Brewster, J. L. (1994). *Onions and other vegetable alliums*. (2nd Edition). Crop production science in horticulture series. ISBN: 9781845933999. 2008.

Burba, L. (1993). Manual de producción de semillas hortícolas. Semillas de ajo. 1ª Ed., *José O. Valderrama, Centro de Información Tecnológica*. Vol. (9). No. 2, pp. 142-163, ISSN: 0716-8756.

Capote, R.A., Fundora, M.Z. y Pérez, D.O. (2000) Estudio de la variabilidad inducida en células y plántulas de cebolla (*Allium cepa* L.) cv Caribe-71 regeneradas *in vitro*. *Elfos scientiae*. Vol. 17, No. 4, (2000), pp. 241-246.

Chávez, C.M, Valenzuela, C.P., Fierros, L.G.A. & Maldonado, N.L.A. (2008). Efecto de métodos y densidades de siembra en la producción de dos variedades de ajo jaspeado en la sierra baja de sonora. XI Congreso Internacional en Ciencias Agrícolas. Universidad Autónoma de Baja California, Mexicali, Baja California. p. 381-385.

Choi, H.S., Kim, K.T., Ahn, Y.K., Kim, D.S. & Woo, J.G. (2003). Analysis of genetic relationships in garlic germoplasm and fertile garlic by RAPD. *Journal of the Korean Society for Horticultural Science*. Vol. 44, No 5, pp. 595-600, ISSN: 0253-6498

Con, L. (2009). Garlic breeding in Cuba. De: http://www.actahort.org/books/433/433_25.htm, 24 sep.

Doyle, J. J. & J. L. Doyle (1990) A rapid total DNA preparation procedure for fresh plant tissue. Focus 12:13-15.

Eom, E. & Lee, D. (1999). Characterization of chromosomal DNA polymorphism in Korean cultivars of *Allium sativum* L. *Journal of Plant Biology*.Vol.42, No.2, pp: 159-167.

FAOSTAT. (2011). FAOSTAT – Crop Statistics. *In:* FAOSTAT URL. Access: 09-12-2011. http://faostat.fao.org/site/567/DesktopDefault.aspx?PageID=567#ancor

Fehér, A., Pasternak, T.P. & Dusits, D. (2003). Transition of somatic plant cells to an embryogenic state. *Plant Cell Tissue and Organ Culture*. Vol. 74, pp. 201-228. ISSN.

Fereol, L., Chovelon, V., Causse, S., Michaux-Ferriere, N & Kahane, R.R. (2002). Evidence of a somatic embryogenesis process for plant regeneration in garlic (*Allium sativum* L.) *Plant Cell Reports*. Vol. 21.

Ferrer, E., Linares, C. & González, J.M. (2000). Efficient transient expression of the β-glucoronidase reporter gene in garlic (*Allium sativum* L.). Vol. 20, No. 8., pp. 869-874, DOI: 10.1051/Agro:200101.

Figliuolo, G., Candido, V., Logozzo, G., Miccolis, V. & Spagnoletti-Zeuli, P.L. (2001). Genetic evaluation of cultivated garlic germplasm (*Allium sativum* L. and *A. ampeloprasum* L.). *Kluwer Academic Publishers*. Vol. 121, pp. 325-234.

García-Lampasana, S., Martínez, L. & Burba, J.L. (2003). Genetic diversity among Argentinean garlic clones (*Allium sativum* L.) using AFLP (Amplified Fragment Length Polymorphism). *Euphytica: Biomedical and Life sciences*.Vol. 132. No. 1, pp. 115-119, DOI: 10.1023/A:1024606004596.

González, A. (2006). "AKUKELI" una nueva variedad de ajo rosado. *Agricultura Técnica*, Vol.66, No.2, pp.210-215.

Haque, M.S., Wada, T. & Hattori, K. (1997) High frequency shoot regeneration and plantlet formation root tip of garlic. *Plant Cell Tissue and Organ Culture*, Vol.51, No.3, pp. 233, DOI:10.1023/A:1005753624853.

Heredia, G.E. & Delgadillo, S. (2002). El ajo en México. Celaya, Gto; México. 37-45 p.

Heredia, Z.A. & Heredia, G.E. (2000). Mejoramiento genético de ajo en el INIFAP. *En: El ajo en México, origen, mejoramiento genético, tecnología de producción.* Libro técnico Núm. 3. México: Secretaría de Agricultura, Ganadería y Desarrollo Rural. Instituto Nacional de Investigaciones Forestales, Agrícolas y Pecuarias. Centro de Investigación Regional Centro. Campo Experimental Bajío, pág. 29-32.

Ipek, M., Ipek, A. & Simon, P.W. (2003). Comparison of AFLPs, RAPD markers, and isozymes for diversity assessment of garlic and detection of putative duplicates in germplasm collections. *J. Amer. Soc. Hort. Sci.* Vol.128, No.2, pp. 246-252.

Ipek, M., Ipek, A. & Simon, P.W. (2006). Sequence homology of polymorphic AFLP markers in garlic (*Allium sativum* L.). *Genome*, Vol. 49, pp. 1246-1255, DOI:10.1139/G06/092

Ipek, M., Ipek, A., Almquist, S.G. & Simon, P.W. (2005). Demonstration of linkage and development of the firs low-density genetic map of garlic, based on AFLP markers. *Theorical and Applied Genetics*, Vol. 110, No.2, pp. 228-236, ISSN: 00405752.

Khar, A., Yadav, R.C., Yadav, N. & Bhutani, R.D. (2005) Transient *gus* expression studies in onion (*Allium cepa* L.) and garlic (*Allium sativum* L.). *Akdeniz Universitesi Ziraat Falultesi* Dergisi.Vol.18, No.3, pp. 301-304.

Kondo, T., Hasegawa, H. & Suzuki, M. (2000). Transformation and regeneration of garlic (*Allium sativum* L.) by *Agrobacterium*-mediated gene transfer. *Plant Cell Reports*, Vol.19, No.10, pp. 989-993. DOI: 10.1007/s002990000222.

Koul, A.K., Gohil, R.N. & Langer, A. (1979). Prospects of breeding improved garlic in the light of its genetic and breeding systems. *Euphytica*, Vol.28, No.2, pp. 457-464. DOI: 10.1007/BF00056605.

Kirk, J.T., Rees, O.H. & Evans, G. (1970). Base composition of nuclear DNA within the genus *Allium*. *Heredity*. Vol. 25, pp. 507-512, DOI: 10.1038/hdy.1970.59

Lagunes, F.E. (2009). Transformación genética de ajo (*Allium sativum* L.) mediante *Agrobacterium tumefaciens*. Tesis de maestría. Colegio de Postgraduados, Campus Montecillo, Montecillo, Texcoco, Edo. de México. p.95.

Levan, A., Fredga, K. & Sandberg, A. (1964). Nomenclature for centromeric possition on chromosomes. *Hereditas*, Vol.52, pp. 201-220, DOI: 10.1111/j.1601-5223.1964.tb01953.x.

Macías, D.R., Grijalva, C.R.L. & Robles, C.F. (2010).). Productividad y calidad de variedades de ajo (*Allium sativum* L.) bajo condiciones desérticas en Caborca, Sonora. *BIOtecnia.* Vol. 12, No.1, (2010), pp.44-54.

Macías, V.L.M. & Maciel, P.L.H. (2009). Mejoramiento genético de diferentes tipos de ajo para la región productora de Aguascalientes. *In*: Memoria X Seminario de Investigación. Universidad Autónoma de Aguascalientes. Aguascalientes, Ags. México. 214-217 p.

Macías, V.L.M., Maciel, P.L.H. & Silos E.H. (2010). Diamante: variedad de ajo tipo California para el Centro Norte de México.INIFAP, CIRNOC, Campo Experimental Pabellón, Pabellón de Arteaga, Ags. México. 40 p. (Folleto técnico No. 41). ISBN: 978-607-425-509-6.

Moriconi, D.N., Burba, J.L. & Izquierdo, J. (1991). Manual de intercambio y propagación de germoplasma de ajo a través de microbulbillos. *FAO Online Catalogues: F02-Plant propagation*. Fiche No. 322397, 45 p. (1991).

McCollum, G.D. (1987). Onion and allies, pp. 186-190. In N.W. Simmonds (ed.) Evolution of crop plants.

Mujica, H., Sanabria, M.E., Mogollón, N. & Perozo, Y. (2008). Formación *in vitro* del bulbo del ajo morado (*Allium sativum* L.). *Revista de la Facultad de Agronomía (LUZ)*. Vol. 25, No. 2, pp. 197-210, ISSN: 1690-9763.

Murashige, T. & Skoog, F. (1962). A revised medium for rapid growth and bioassay with tobacco tissue cultures. *Physiologia Plantarum*, Vol. 15, No 43, pp. 437-497, DOI: 10.1111/j.1399-3054.1962.tb08052.x . Article first published online: 28 APR 2006.

Nei, M. & Li, W. (1979). Mathematical model for studying genetic variation in terms of restriction endonucleases. *Proc. Natl. Acad. Sci. USA*. Vol. 76, No. 10, pp. 5256-5273. (august 1979).

Novák, F.J., Havel, L. & Dolezel, J. (1983). *Allium In* Crop Species. Hand book of Plant Cell Cultures. Techniques and Applications. Inst. Exp. Botany. Czechoslovak Acad. Sci. Praha. 419-455 p.

Panzera, F., Gimenez-Abian, M.I., López-Saez, J.F., Jiménez-Martín, G., Cuadrado, A., Shaw, P.J., Beven, A.F., Canovas, J.L. & De la Torre, C. (1996). Nucleolar organizer expresión in *Allium cepa* L. chromosomes. *Chromosoma*, Vol. 5, No.1, pp.12-19. DOI: 10.1007/BF02510034.

Pardo, A., Hernández, A. & Méndez, N. (2009). Caracterización molecular de siete clones de ajo (*Allium sativum* L.) mediante la técnica RAPD. *Bioagro*. Vol. 21, No. 2, pp. 81-86, ISSN: 1316-3361.

Paredes C. M., Becerra V. & González A. M. I., (2008). Low genetic diversity among garlic (*Allium sativum* L.) accessions detected using random amplified polymorphic DNA (RAPD). Chilean Journal of Agricultural Research, Vol. 68, No. 1:3-12.

Peschke, V. & Phillips, R.L. (1992). Genetic implications of somaclonal variation in plants. *Advances in Genetics*, Vol. 30, pp. 41-45, doi:10.1016/S0065-2660(08)60318-1.

Quintana-Sierra, M. E., A. Robledo-Paz., A. Santacruz-Varela., M. A. Gutiérrez-Espinosa., G. Carrillo-Castañeda & J. L. Cabrera-Ponce, (2005). Regeneración *in vitro* de plantas de cebolla (*Allium cepa* L.). Agrociencia 39: 647-655.

Rabinowitch, H. D. & Brewster J. M., (1990). Onions and Allied Crops. Botany, Physiology and Genetics. Vol. I CRC press. Inc. Boca Raton, Florida. U.S.A. 273 p.

Robledo-Paz, A., Villalobos-Arámbula, V.M. & Jofre-Garfias, A.E. (2000). Efficient plant regeneration of garlic (*Allium sativum* L.) by root tip culture. *In Vitro Cellular & Developmental Biology*, Vol. 36, No. 5, pp. 416-419, DOI: 10.1007/s11627-000-0075-6

Robles P. J., Armenta C. R. A., & Valenzuela C., (2006). México en el Contexto Global de la Producción de Ajo. En Memorias, Seminario Técnico: Tecnología para la producción de ajo en la sierra de Sonora. Universidad de Sonora-INIFAP-Fundación Produce Sonora. p. 7-14.

Molina, M.L.G. & Rosales, L.F.U. (2007). Diversidad genética de poblaciones de ajo (*Allium sativum* L.) cultivadas en Guatemala, definida por marcadores de ADN. *Agronomía Mesoamericana*, Vol. 18, No. 1, pp. 85-92, ISSN: 1021-7444

Sato, S., Hizume, M. & Kawamura, S. (1980). Relationship between secondary constrictions and nucleolus organizing regions in *Allium sativum*. *Protoplasma*, Vol. 105, No. 1-2, pp. 75-85, DOI: 10.1007/BF01279851

Shasany A. K., Ahirwar O. P., Kumar S. & Khanuja S. P., (2000). RAPD analysis of phenotypic diversity in the Indian garlic collection. *Journal of Medicinal and Aromatic Plant Sciences*. Abstr. 22(1B): 586-592.

SIAP. 2011. (http://reportes.siap.gob.mx/Agricola_siap/ResumenProducto.do)

Spangenberg, G. (1997). Transformation methods and analysis of gene expression in transgenic plants. ICGEB-Course. Embrapa. Brasilia, Brazil. 59-65 p.

Trejo P. P. (2006). Presentación, II Foro Nacional del Ajo. Memorias. Gobierno de Zacatecas: INIFAP, Fundación Produce, Zacatecas, SAGARPA, FIRA, Consejo Estatal de Productores de Ajo de Zacatecas A.C. Zacatecas, Zacatecas. México. P. 9-13.

Shigyo, M., Wako, T., Kojima, A. Yamauchi, N. & Tashiro, Y. (2003). Transmission of allien chromosomes from selfed progenies of a complete set of *Allium* monosomic additions: the development of a reliable method for the maintenance of a monosomic addition set. Genome, Vol. 46, No. 6, pp. 1098-1103, DOI: 10.1139/G03-075

Van Heusden, A.W., Van Ooijen, J.W., Vrielink, R., Ginkel, V., Verbeek, W.H.J., Wietsma, W.A. & Kik, C. (2000). A genetic map interspecific cross in *Allium* based on amplified fragment length polymorphism (AFLP) markers. *TAG Theorical and Applied Genetics*, Vol. 100, No. 1, pp. 118-126, DOI: 10.1007/s001220050017

Van der Valk, P., Scholten, O. E., Verstappen, F., Jansen, R.C. & Dons, J.J.M. (1992). High requency somatic embryogenesis and plant regeneration from zygotic embryo-derived callus cultures of three *Allium* species. *Plant Cell Tissue and Organ Culture*, Vol. 30, No. 3, pp. 181-191, DOI: 10.1007/BF00040020

Verma, S. C. & R. K. Mittal, (1978). Chromosome variation in the common garlic. Cytologia. 43: 383-396.

Xu, P., Liu, H., Gao, Z., Srinives, P. & Yang, C. (2001). Genetic identification of garlic cultivars and lines by using RAPD assays. *ISHS Acta Horticulturae* 555: II *International Symposium on Edible Alliaceae*, pp. 213-220.

Zheng, S., Henken, B., Sofiari, E., Jacobsen, E., Krens, F.A. & Kik, C. (1998). Factors influencing induction, propagation and regeneration of mature zygotic embryo-derived callus from *Allium cepa*. *Plant Cell Tissue and Organ Culture*, Vol. 53, No.2, pp. 99-105, DOI: 10.1023/A: 1006034623942

Permissions

The contributors of this book come from diverse backgrounds, making this book a truly international effort. This book will bring forth new frontiers with its revolutionizing research information and detailed analysis of the nascent developments around the world.

We would like to thank Mahmut Caliskan, for lending his expertise to make the book truly unique. He has played a crucial role in the development of this book. Without his invaluable contribution this book wouldn't have been possible. He has made vital efforts to compile up to date information on the varied aspects of this subject to make this book a valuable addition to the collection of many professionals and students.

This book was conceptualized with the vision of imparting up-to-date information and advanced data in this field. To ensure the same, a matchless editorial board was set up. Every individual on the board went through rigorous rounds of assessment to prove their worth. After which they invested a large part of their time researching and compiling the most relevant data for our readers. Conferences and sessions were held from time to time between the editorial board and the contributing authors to present the data in the most comprehensible form. The editorial team has worked tirelessly to provide valuable and valid information to help people across the globe.

Every chapter published in this book has been scrutinized by our experts. Their significance has been extensively debated. The topics covered herein carry significant findings which will fuel the growth of the discipline. They may even be implemented as practical applications or may be referred to as a beginning point for another development. Chapters in this book were first published by InTech; hereby published with permission under the Creative Commons Attribution License or equivalent.

The editorial board has been involved in producing this book since its inception. They have spent rigorous hours researching and exploring the diverse topics which have resulted in the successful publishing of this book. They have passed on their knowledge of decades through this book. To expedite this challenging task, the publisher supported the team at every step. A small team of assistant editors was also appointed to further simplify the editing procedure and attain best results for the readers.

Our editorial team has been hand-picked from every corner of the world. Their multi-ethnicity adds dynamic inputs to the discussions which result in innovative outcomes. These outcomes are then further discussed with the researchers and contributors who give their valuable feedback and opinion regarding the same. The feedback is then collaborated with the researches and they are edited in a comprehensive manner to aid the understanding of the subject.

Apart from the editorial board, the designing team has also invested a significant amount of their time in understanding the subject and creating the most relevant covers. They scrutinized every image to scout for the most suitable representation of the subject and create an appropriate cover for the book.

The publishing team has been involved in this book since its early stages. They were actively engaged in every process, be it collecting the data, connecting with the contributors or procuring relevant information. The team has been an ardent support to the editorial, designing and production team. Their endless efforts to recruit the best for this project, has resulted in the accomplishment of this book. They are a veteran in the field of academics and their pool of knowledge is as vast as their experience in printing. Their expertise and guidance has proved useful at every step. Their uncompromising quality standards have made this book an exceptional effort. Their encouragement from time to time has been an inspiration for everyone.

The publisher and the editorial board hope that this book will prove to be a valuable piece of knowledge for researchers, students, practitioners and scholars across the globe.

List of Contributors

Dobrinka Atanasova, Nikolay Tsenov and Ivan Todorov
Dobrudzha Agricultural Institute, General Toshevo, Bulgaria

Lai Van Lam, Tran Thanh, Le Thi Thuy Trang, Vu Van Truong, Huynh Bao Lam and Le Mau Tuy
Rubber Research Institute of Vietnam, Ho Chi Minh City, Vietnam

Zabardast T. Buriev, Shukhrat E. Shermatov, Alisher A. Abdullaev, Khurshid Urmonov, Fakhriddin
Kushanov, Sharof S. Egamberdiev, Umid Shapulatov, Abdusttor Abdukarimov and Ibrokhim Y. Abdurakhmonov
Center of Genomic Technologies, Institute of Genetics and Plant Experimental Biology Academy of Sciences of Uzbekistan, Yuqori Yuz, Qibray Region, Tashkent, Uzbekistan

Sukumar Saha and Johnnie N. Jenkins
United States Department of Agriculture-Agriculture Research Service, Crop Science Research Laboratory, Starkville, Mississippi State, USA

Russell J. Kohel and John Z. Yu
United States Department of Agriculture-Agriculture Research Service, Crop Germplasm Research Unit, College Station, Texas, USA

Alan E. Pepper
Department of Biology, Texas A&M University, College Station, Texas, USA

Siva P. Kumpatla
Department of Biotechnology Regulatory Sciences, Dow AgroSciences LLC, Indianapolis, Indiana, USA

Mauricio Ulloa
United States Department of Agriculture-Agriculture Research Service, Western Integrated Cropping Systems, USA

C. O. Aremu
Department of Crop Production and Soil Science Ladoke Akintola University of Technology, Ogbomoso, Oyo state Landmark University, Omu-Aran, Nigeria

Danielle Cristina Gregorio da Silva and Cristiano Medri
Universidade Estadual do Norte do Paraná, Brazil

Mayra Costa da Cruz Gallo de Carvalho
Empresa Brasileira de Pesquisa Agropecuária, Brazil

Moacyr Eurípedes Medri, Claudete de Fátima Ruas, Eduardo Augusto Ruas and Paulo Maurício Ruas
Universidade Estadual de Londrina, Brazil

Elena Artyukova, Marina Kozyrenko, Olga Koren, Alla Kholina, Olga Nakonechnaya and Yuri Zhuravlev
Institute of Biology and Soil Science, Far East Branch of Russian Academy of Sciences, Russia

Yamini Kashimshetty and Steven H. Rogstad
Department of Biological Sciences, USA

Melanie Simkins
Department of Environmental Studies, USA

Stephan Pelikan
Department of Mathematical Sciences, University of Cincinnati, OH, USA

Dagmar Janovská
Department of Gene Bank, Crop Research Institute, Czech Republic

Petra Hlásná Čepková
Department of Crop Sciences and Agroforestry in Tropics and Subtropics, Institute of Tropics and Subtropics, Czech University of Life Sciences Prague, Czech Republic

Mária Džunková
Department of Genomics and Health, Centre for Public Health Research (CSISP), Valencia, Spain

Héctor Silos Espino, Flora San Juan Hernández, Olivio Hernández Hernández, Darío Silva Bautista, Alan Roy Macías Ávila, Francisco Nieto Muñoz, Luis L. Valera Montero and Silvia Flores Benítez
Biotechnology Applied Laboratory, Instituto Tecnológico El Llano, Aguascalientes (ITEL), km. 18 Carr. Aguascalientes-San Luís Potosí, México

Luis Martín Macías Valdez
Instituto Nacional de Investigaciones Forestales, Agrícolas y Pecuarias. Campo Experimental, Pabellón de, Arteaga, km 32.5 carr. Aguascalientes-Zacatecas, México

Tarsicio Corona Torres
Cytogenetic Laboratory "Maestra Czeslawa Prywer Linzbarka", Colegio de Postgraduados, km. 36.5, carr., México-Texcoco Montecillo, Texcoco, Estado de México, México

Mario Leonel Quezada Parga
Fundación Produce Aguascalientes A. C. Av. Universidad 604 Interior 6 y 7, Fracc. Unidad Ganadera, Aguascalientes Ags., México

Juan Florencio Gómez Leyva
Molecular Biology Laboratory, Instituto Tecnológico de Tlajomulco, Tlajomulco de Zúñiga, Jalisco, Mexico

Printed in the USA
CPSIA information can be obtained
at www.ICGtesting.com
JSHW011358221024
72173JS00003B/332

9 781632 392534